Shadow

and Psychic Phenomena

A scientist casts new light on psychic phenomena, such as clairvoyance, telepathy, and 'out of the body experiences' in this book. He presents an exciting new theory which explains such phenomena, linking the recently discovered 'Shadow Matter' world of physics with parapsychology. It replaces notions of the occult by important new ideas that are figuring in physics (even in recent television programmes).

The book contains case histories, showing how this new theory could account for telepathy, clairvoyance, 'out of the body experiences,' and apparitions of the living and the dead in terms of Shadow Matter. It also explains how Shadow Matter theory could account for the survival of the human personality after the death of the body.

In *Shadow Matter and Psychic Phenomena*, the author builds on theories until now discussed only in academic journals. This theory was first published in brief (now dated) outline in the journal *Inquiry* in 1988. It develops the concept of the Shadow Matter world introduced in the journal *Nature* in 1985.

* The first book for the non-specialist to integrate parapsychology with mainline science
* Shows succinctly but non-technically how the most modern theoretical physics of Superstrings and Shadow Matter could explain most or all psychic phenomena
* Includes over eighty fascinating case histories.

In Memory of
Dr Kossy Strauss
First Technical Director of (later) Foseco Minsep
and in Honour of his widow
Mrs Bertha Strauss

Shadow Matter and Psychic Phenomena

A scientific investigation into psychic phenomena and possible survival of the human personality after bodily death

by Gerhard D Wassermann Ph.D, FIMA
(Formerly Reader in the Theory and Philosophy of Biology,
University of Newcastle upon Tyne)

Mandrake of Oxford

Note about the Author
After graduating with first class honours in Mathematics from Queen Mary College, University of London, Gerhard Wassermann obtained a PhD in quantum mechanics at the same university where at Birkbeck College he also studied Biology. Then he joined Professor Herbert Fröhlich FRS in the Bristol Physics Department of Sir Nevill Mott FRS. He later became a lecturer in Applied Mathematics at what is now the University of Newcastle upon Tyne, teaching at undergraduate and postgraduate levels. His research interests shifted to theoretical biology and philosophy of science and philosophy of mind. He became Reader in the Theory and Philosophy of Biology, and also Associate Editor of the *Bulletin of Mathematical Biology*. He has also twice been a visiting professor abroad. He has 50 publications, including three books, to his credit.

Previous Books by this Author include:
Brains and Reasoning (London, Macmillan, 1974)

A catalogue record (CIP) for this book is available from the British Library
ISBN 1869928-326

Mandrake *of Oxford*, PO Box 250, OXFORD, ox1 1ap Britain

Contents

Preface

This book tries to provide the beginning of a new understanding of psychic phenomena. To achieve this it attempts to link the psychic world with the world of Shadow Matter. Although borrowed from the realm of modern theoretical physics, Shadow Matter might also appeal to students of the occult, for reasons to be given. Understanding what Shadow Matter can achieve requires in the present book next to no physics and no mathematics. I shall try to explain, in largely new, but simple terms, possible ways in which psychic phenomena could come about. In particular, I shall try to show how life after death could be possible, and I shall demolish old arguments which dispute this. A few of my ideas were published in very condensed form, not addressed to laymen, in the journal *Inquiry* (Wassermann, 1988). Here I present a much expanded version which should make my views accessible to the educated public at large. The present formulation, addressed to general readers, contains also numerous ideas not dealt with in my 1988 paper, and explains some important psychic phenomena which I did not consider earlier, notably 'out of the body experiences'. By understanding how psychic phenomena could work, people may, at last, accept these phenomena as ingredients of normal science, rather than look on them as inexplicable curiosities. Indeed, if, as I maintain, psychic phenomena are linked to theoretical physics, then, via physics, they are linked to science.

As noted, I believe that the key to an understanding of the 'machinery' of psychic phenomena is *Shadow Matter*. This was introduced in 1985 in the important science journal *Nature* and is assumed to be a type of matter very different from ordinary matter which is made up of familiar constituents such as atoms and molecules. Although different from ordinary matter, Shadow Matter can, according to physicists, interact gravitationally with ordinary matter. By making a few, very simple, assumptions about Shadow Matter it is possible to show in considerable detail that Shadow Matter, together with ordinary matter, could explain most, or all, psychic phenomena. A pathological prejudice against psychic phenomena has been fostered by certain people, who frown on the unexplained, and, perhaps, this book may help to sweep aside this ill-founded prejudice.

When Shadow Matter was first introduced into physics by Kolb *et al* (1985) it was intended, exclusively, for cosmologists, i.e. people who theorize about the structure and behaviour of the universe. From 1987 onwards, however, I was increasingly struck by the idea that Shadow Matter, jointly with ordinary matter, could generate also telepathy, clairvoyance, 'out of the body experiences' and other psychic phenomena. In addition, Shadow Matter could ensure survival of the human personality after death of the ordinary matter body (Wassermann, 1988). How this, and much else, could come about will be explained in this book.

Some philosophers, and many religious people, have argued that man consists of an ordinary matter body and an immaterial, immortal, soul, and that body and soul interact during life. Likewise, occultists (e.g. Rudolf Steiner, 1969) have invoked entities, such as the 'astral body,' in addition to the 'ordinary matter body.' Instead of this, I shall argue that man consists of an ordinary matter body and, in addition, a Shadow Matter body which includes a Shadow Matter brain. (Shadow Matter could be an immensely light substance.) Unlike the 'astral body,' the Shadow Matter body is assumed to have specified physical properties. After death of the ordinary matter body the Shadow Matter body and its Shadow Matter brain could live on, possibly indefinitely.* It might be tempting to interpret the Shadow Matter brain as a material soul. In fact, materialistic philosophers of ancient Greece believed already that souls consist of matter (see Rist, 1972; Long, 1974). Likewise, many occultists might be tempted to identify one of their assumed occult 'bodies' with the present Shadow Matter body. Irrespective of whether this is the case, the ideology of this book and occultism share the view that there exists more than one kind of human body, namely, according to my view an ordinary matter body and a Shadow Matter body. I trust that some readers will find these new ideas about the supposed potentialities of the world of Shadow Matter as exciting as I did in writing this book.

Gerhard D. Wassermann
Newcastle upon Tyne 1993

* As an aside for readers who know some physics. For all we know, the second law of thermodynamics (familiar to physicists) might not apply to Shadow Matter.

Introduction

I wish to cast new light on the nature of psychic phenomena. This book was written with missionary zeal. It contains many new or recent ideas of mine, many appearing here for the first time. Primarily this work is meant for educated laymen and requires no specialized knowledge, not even the most elementary mathematics. Besides such general readers, the book may also interest specialists in various fields in schools, colleges and universities and it may arouse the curiosity of religious people as much as that of agnostics and philosophers. For instance, at Edinburgh University there exists now a Koestler chair for the study of psychic phenomena. Currently it is occupied by Professor Robert Morris. This shows that these topics are also of concern to various academics and not only to laymen. At any rate, many people would like to know what could happen to them (if anything!) after they *depart* from this world. They would also like to know how telepathy, clairvoyance and other psychic phenomena work.

My principal aims are two-fold. First, I wish to present a new, coherent, theory, which explains in simple terms, that can be understood by general readers, how psychic phenomena could work. This new theory, by making sense of these phenomena, could help to dispel much unjustified scepticism. Second, the theory casts also new light on the possibility of survival of the human personality after death of the human body. Such a unified theory existed up to now only in embryonic form (Wassermann, 1988) which was published in the journal *Inquiry*. Consequently many psychic phenomena have hitherto existed still as isolated bits and pieces, curiosities, which sceptics dismissed out of hand. These sceptics insist still that these phenomena cannot exist because they are *impossible*, or that they defy the laws of physics. I believe that it is a complete waste of time to argue with the more dogmatic of these sceptics. They are as irrational as those people who, in Nazi Germany, argued that the special theory of relativity must be wrong because it was discovered by Einstein, a Jew. I have met also totally irrational people of some standing in academic positions, who maintained that psychic phenomena simply cannot and do not occur, because they say so. If you tell such people that psychic phenomena can be explained rationally in terms of a coherent theory which, indirectly, links them with theoretical physics, then they change the argument. They will claim that such explanations must be wrong because the phenomena do not exist and, hence, need not, and cannot, be explained.

One must console oneself that this kind of dogmatism by irrational people of some standing has rarely retarded the progress of science. Such people exist, and have existed, in many branches of science and among various philosophers. Tyrrell (1948, p. 259) pointed out that:

'That great philosopher Bacon,' writes Professor Macneile Dixon, 'could not to the last believe that the earth revolved round the sun. The facts were too solidly opposed to such a fancy. It was incredible. The diamond appears the acme of stability, it is in fact a whirlpool of furious motion [on the atomic and sub-atomic scales]. Who could believe it? What is credible? Only the familiar. When the news of the invention of the telephone was reported to Professor Tait, of Edinburgh, he said "It is all humbug, for such a discovery is physically impossible." When the Abbé Moignon first showed Edison's phonograph to the Paris Academy of Sciences, all the men of science present declared it impossible to reproduce the human voice by means of a contraption and the Abbé was accused, Sir William Barrett tells us, of having a ventriloquist concealed beneath the table. The thing was unbelievable.'[1]

I suppose that many people, with hindsight, regard the above views of Bacon and Tait as outrageous pieces of scientific prejudice. Yet, the opinions of Emeritus Professor of Psychology Max Hammerton (see 1983), on the topic of psychic phenomena, strike me as no more persuasive than, say, Tait's outburst about the physical impossibility of the telephone. The reader can find a more extensive treatment of the topic of hyperscepticism and scientific prejudice concerning psychic phenomena in section 1.11.

While my new unified theory will not shift the views of the pathologically prejudiced, it might appeal to more rational people. Integrating psychic phenomena within a coherent theoretical framework could help to fashion these phenomena into a genuine science. Surely, a mountain of facts, without a theory to make sense of them, is no science. More important, my theory links psychic phenomena with recent major discoveries in theoretical physics. Psychic phenomena could thereby become a part of orthodox science. This could be so irrespective of what extremist sceptics, like James Randi, a conjurer whose scientific credentials I do not know, have to say. The link between psychic phenomena and theoretical physics comes via *string* physics. String theory was introduced into physics relatively recently by M. B. Green FRS (see 1985) and others to be mentioned. Professor Green received his Fellowship of the Royal Society (London) for his important contributions to string theory. What precisely string theory is concerned with does not matter at this stage. Let me just say that it deals with the way that 'elementary particles,' such as electrons and quarks, are composed, and with much else. There are numerous varieties of string theories possible within string physics. But exactly one kind of string theory led to an important conclusion. It suggested that in addition to the familiar 'ordinary matter' of which we, and all objects around us are primarily composed, there could exist another kind of matter. This other matter, introduced into physics by Kolb *et al.* (1985), has been called *Shadow Matter*.[2]

If Shadow Matter does, in fact, exist, then it can be shown, as will be done in most of this book, that, provided Shadow Matter has a few simple properties, Shadow Matter, together with ordinary matter, could explain most or all psychic phenomena. To grasp the relationship between Shadow Matter and psychic phenomena no previous knowledge of mathematics or physics is required. I am, after all, not trying to demonstrate that Shadow Matter could be a mathematical consequence of string theory. Such a demonstration does, indeed, exist, and is mathematically very intricate (Kolb *et al* 1985). Its conclusions are accepted here as known and without mathematical proof. Thus, the primary plausibility of the possible existence of Shadow Matter comes from string theory in physics. I believe that the explanations of psychic phenomena, given in this book in terms of Shadow Matter theory could provide a secondary plausibility for the existence of Shadow Matter. Thus, psychic phenomena 'could provide, perhaps, the best *indirect* evidence for the existence of Shadow Matter' (Wassermann, 1988).

Before I discuss, in anticipatory outline, what part Shadow Matter could play in parapsychology, I shall turn to other matters first. Let me inform readers, who, perhaps, do not know, and remind those who know, what psychic phenomena comprise. Psychic phenomena, usually called *psi-phenomena* for short, include: telepathy, clairvoyance, apparitions (of the dead and the living), precognition (i.e. foreknowledge), 'out of the body experiences,' called OBEs for short, and synchronicity experiences, among others. Telepathy and clairvoyance are often collectively subsumed under the heading of extra-sensory perception, or ESP for short. People who study psi-phenomena are known either as parapsychologists or as psychical researchers. Attempts to define particular psi-phenomena involve usually already some kind of theory concerning the phenomena. For instance, the 5th edition of the *Concise Oxford Dictionary* defines telepathy as 'Action of one mind on another at a distance through emotional influence without communication through senses.' This mind-orientated definition, however, is far removed from the kind of science-related, matter-orientated interpretation of telepathy presented later in this book. Similar remarks apply to the *Concise Oxford Dictionary* definition of clairvoyance as the 'Faculty of seeing mentally what is happening or exists out of sight.' I shall avoid, therefore, formal definitions of psi-phenomena. Instead, my meanings of particular psi-phenomena will surface in relation to typical case histories and in connection with my theory of psi-phenomena. In fact, the book contains more than seventy, richly structured, case histories of spontaneous psi-phenomena. These should give readers an idea of the intricacy of these phenomena. They will be used also to demonstrate how my theory can be applied to typical spontaneous cases.

It is often said that mechanistic materialism is incompatible with psi-phenomena. Since many scientists (including the author) are mechanistic materialists, many of these scientists (but not the author) have rejected, accordingly, the findings of parapsychologists. Often these mechanistic materialistic scientists simply could not be bothered to examine these findings, but rejected them just the same. Mechanistic materialism must not be confused with Marxist 'dialectical materialism' and has nothing to do with politics. Mechanistic materialism asserts simply that all properties of the universe are properties of matter, and that mentality is also a property of matter. Perhaps the most startling result of this book is that survival of the human personality after bodily death is fully compatible with mechanistic materialism and that survival is possible (but not logically necessitated by any known facts). This conclusion seems to go hand in hand with my interpretations, by means of my theory, of a number of reported 'out of the body experiences' (OBEs).

In typical OBEs people claim to have left their ordinary matter body while alive. They claim also to have seen this body from the outside while residing away from this body. While away from that body they sometimes visited also neighbouring localities before reuniting with their ordinary matter body. If, apparently, people can leave their ordinary matter bodies, while alive, during an OBE, then it would not be surprising if they leave that body after death, and continue to exist independently of that body. Yet, how is this possible?[3] The answer to this question takes up most of this book. But I shall anticipate it here in brief outline. This brings me back to Shadow Matter.

My theory assumes that all 'ordinary matter,' whether solid, liquid or gas, could *bind* accurately matching Shadow Matter. Thus, each specific ordinary matter constituent could, like a very specific key matches a very specific lock, match a complementary, equally specific, Shadow Matter constituent. The ordinary matter *key* and its matching Shadow Matter *lock* could become bound to each other. According to Kolb *et al* (1985) ordinary matter and Shadow Matter can interact only gravitationally, i.e. by the force of gravity. It is known to physicists that the gravitational force is extremely weak compared to other known forces that produce interactions between ordinary matter. Nevertheless, a constituent of ordinary matter could attract gravitationally strongly at close range, a matching *complementary* constituent of Shadow Matter. The gravitational interaction could become very strong as the Shadow Matter constituent interlocks with the ordinary matter constituent as they come close together. In this way the matching constituents could attract each other sufficiently strongly, gravitationally, to get stuck to each other, i.e. *to become bonded* (as physicists and chemists would say). (The bond involved here is a gravitational one, whereas a chemical bond, say, between two matching atoms of ordinary matter is mediated in quite different ways.)

Part of the theory technically stated

To apply the preceding ideas, let us look first at ordinary matter. This is generally composed of atoms, ions, molecules and free electrons. Each atom consists of an atom-specific nucleus surrounded by an atom-specific number of electrons. The nucleus of any atom is composed of atom-specific numbers of two kinds of elementary particles known as up-quarks and down-quarks respectively (which are mainly held together by gluons).

My theory assumes that each up-quark could become bound to a Shadow Matter up-quark. Likewise each down-quark could become bound to a Shadow Matter down-quark. Accordingly each atomic nucleus could be bonded (maximally) to as many

---

Shadow Matter quarks (called squarks for short) as it contains ordinary matter quarks. The squarks bound to a nucleus could interact by Shadow Matter forces to form a Shadow Matter nucleus (called a snucleus) which is bound to the nucleus, as explained, i.e. each quark binding a squark (optimally). Each of the electrons surrounding an atomic nucleus is assumed to be able to bind a Shadow Matter electron (called a selectron for short). The snucleus and selectrons bound to an atom could interact with each other by Shadow Matter forces and, thereby, form a Shadow Matter atom; (here called satom). In this way each atom could be interacting (gravitationally) with, and be bound to, a specific satom.

Similarly, ordinary matter atoms which are linked up to form molecules could have their companion satoms linked up (by Shadow Matter forces) to form smolecules and so forth.

Part of the theory simply put

In this way all ordinary matter objects, whether as large as an elephant (or larger), or as minute as an electron, could be accompanied by precisely matching Shadow Matter objects (also called Shadow Matter Models) which are bound to them. Thus, to the elephant there could be bound a Shadow Matter selephant such that to each atom of the elephant there is bound a corresponding, matching, satom of the selephant. And what applies to the elephant could apply to innumerable animate and inanimate objects, including human beings. Thus each ordinary matter human body, while alive, could be bound to an atom-for-satom matching Shadow Matter body which represents the human body and all its parts, in *shadowy guise.*

Going on in this way, we can see that, according to this theory, the whole of planet Earth, including all constituents of its interior and all things that exist on its surface and in its atmosphere, could be accompanied by a huge Shadow Matter planet Earth, where each atom of one is bound to each matching satom of the other (and similar remarks apply to ions etc). The same story could apply with appropriate changes to the sun and to all objects of all galaxies etc. In addition my theory assumes that there exists a vast amount of *free* Shadow Matter (which is not bound to ordinary matter) in the universe.

According to the preceding assumptions, it was seen that to our ordinary matter bodies there could correspond Shadow Matter bodies bound to our ordinary matter bodies. In particular, to our ordinary matter brains there could correspond Shadow Matter brains, which are bound to the ordinary matter brains. To each ordinary matter brain atom there would correspond a Shadow Matter brain satom, which is bound to the ordinary matter brain atom. I assume also, contrary to Kolb *et al* (1985), that the weight of each Shadow Matter object is minute compared to the weight of the matching ordinary matter object to which it binds. (I have used weight here, instead of mass, so as not to confuse people who know no physics.)

After this detour let me return to 'out of the body experiences' (OBEs) and life after death. In OBEs most of people's Shadow Matter bodies, including their Shadow Matter brains, could become detached temporarily from their ordinary matter bodies.

The Shadow Matter bodies with their Shadow Matter brains could then move about in space. (During this separation the ordinary matter body and the Shadow Matter body may remain linked by a thin, highly extensible, cord, which has been reported by several people who had OBEs. I am assuming that this cord consists of Shadow Matter and forms part of the Shadow Matter body.)

Some people, notably some who have not studied case histories of OBEs very carefully, have dismissed the possibility that in an OBE some entity separates from the ordinary matter body. Instead they argued that OBEs are entirely due to hallucinations. Others, including myself, have not accepted the 'hallucination only' assumption. I shall argue, repeatedly, that the latter assumption, apparently, cannot account for numerous findings concerning OBEs. These findings are too numerous, and the rebuttal of the 'hallucination only' assumption takes too much space to be discussed at this stage, and will be postponed until later in the book.

I am coming now to the most unconventional assumption of the new theory, not present in my 1988 paper. The conventional view of physiologists, medical people, biologists and most laymen is that the *ordinary matter brain* is the system that provides the physical basis for storing memory traces and for constructing and representing thoughts and feelings.[4] It is also believed to be the system for perceiving the world three-dimensionally and in colour etc. At first sight this view seems to agree with the findings of people who have studied the possible ordinary matter brain basis of memories. These findings showed clearly that the ordinary matter brain is involved, normally, in quite specific ways in the formation of memory traces.[5] (Many brain-scientists favour the view that learning involves changes of particularly located junctions of nerve cells, the so-called synapses. It is thought that appropriately distributed patterns of changed synapses could represent particular memory traces, within the ordinary matter brain. This could be quite correct.) But people can also form memories during 'out of the body experiences' (OBEs). After such an experience they can remember what happened to them during the experience. Moreover, the studies that relate to the ordinary matter brain do not show that the ordinary matter brain is the necessary material basis for representing memories, for constructing and representing thoughts and feelings etc.

The occurrences of changed synapses during formation of new memory traces and the observations of OBEs would still be compatible with the following major assumption made here. I assume that normally and during OBEs, the *Shadow Matter brain, and not the ordinary matter brain, is the physical system which stores and represents memory traces, and which produces and represents thoughts, feelings and perceptions*. According to this assumption, normally, when the Shadow Matter body is assumed to be bound to the ordinary matter body, the ordinary matter brain and nervous system are assumed to mediate between the environment and the Shadow Matter brain. The Shadow Matter brain, while bound to the ordinary matter brain, could communicate with the latter as follows. According to Kolb *et al* (1985) the interaction between Shadow Matter and ordinary matter is assumed to be by gravity. Hence, the ordinary matter brain could communicate with the Shadow Matter brain by minute *packets* of gravitational energy called gravitational quanta or gravitons for short.[6] Also, by means of gravitons, the Shadow Matter brain could trigger critically set motor nerve cells of the ordinary matter brain, thereby initiating nerve impulses that could lead to behaviour e.g. human speech. Hence, thoughts and feelings, as represented by the Shadow Matter brain, could lead to human actions.

According to that present view, when, during an 'out of the body experience,' or as a result of tissue injury, the Shadow Matter body and its Shadow Matter brain separate partially or completely from the ordinary matter body, then the following could happen. When separating, the Shadow Matter brain could retain its capacity to see three-dimensionally and in colour and the ability to think, feel and form memory traces, and could retain the memory traces formed before separation. But how can people 'see' their environment during an 'out of the body experience?' Normally, in addition to their ordinary matter eyes, and bound to them, people could be expected to have Shadow Matter eyes. These Shadow Matter eyes would form part of the Shadow Matter body, already assumed, and would be linked by Shadow Matter nerve tracks to the Shadow Matter brain. In normal vision, which, by assumption is mediated by the ordinary matter eyes, optic nerve and ordinary matter brain to the Shadow Matter brain, the Shadow Matter eyes, because of tight bonding to the ordinary matter eyes, may not be receptive, most of the time, to appropriate environmental Shadow Matter signals for vision. During an 'out of the body experience,' however, when by assumption the Shadow Matter body becomes detached from the ordinary matter body, the Shadow Matter eyes would also become detached from the ordinary matter eyes. When thus detached, the Shadow Matter eyes could become responsive to environmental visual Shadow Matter signals. These signals will be called sphotons, and correspond to the ordinary matter photons (i.e. light particles). Sphotons could then activate the Shadow Matter eyes which, in turn activate the Shadow Matter nervous system of the Shadow Matter body. This could lead to activation of that part of the Shadow Matter brain that, in normal vision, leads to visual representations.

When we die our ordinary matter body and its brain decay or are cremated, etc. But the previously attached Shadow Matter body, including its Shadow Matter brain could persist stably and, hence, survive. Thus, the Shadow Matter brain, with all its stored memories, its capacity to see, think and feel etc., could survive, possibly indefinitely. (See my remarks concerning Shadow Matter and the Second Law of Thermodynamics below.) Accordingly, Shadow Matter theory, as presented in this book, could link closely the possibility of survival of the human personality after death of the ordinary matter body with 'out of the body experiences,' which thousands of people have reported. Possibly, the experience of 'survival' could closely resemble that of an 'out of the body experience.'

The preceding very brief survey will be expanded much in the following chapters. But it indicates already how, in terms of the new theory, 'out of the body experiences' (OBEs) and survival after death could be explained, in closely similar terms, by means of Shadow Matter. Also, in terms of this theory, survival becomes a distinct possibility. This, however, is not the end of the story. My theory assumes (see p. 12) that all ordinary matter, and not just human bodies, can bind Shadow Matter. It assumes also that Shadow Matter, bound to ordinary matter can emit, and interact with, Shadow Matter communication devices. These devices allow very long-distance communication between Shadow Matter, bound to far apart objects, to take place. This kind of communication, as I shall show, avoids the notorious 'inverse square law' difficulty, which has impeded all previous communication theories of psi-phenomena. In fact, telepathy and clairvoyance will be explained in closely similar terms by means of the Shadow Matter communication theory. I shall show also that collective (and other) apparitions, precognition and the hitherto obscure phenomenon of 'object reading' can be explained in terms of Shadow Matter communication theory. My 1988 paper contained a very condensed version of a part of this comprehensive new theory of psi-phenomena but did not deal adequately with clairvoyance and did not deal with OBEs at all. The new theory provides within its explanatory framework also a new partial theory of hypnotic states.

This book liquidates, implicitly, the claims of those who assert that parapsychology undermines materialism. It suggests that psi-phenomena, and even possible survival of the Shadow Matter body after death of the ordinary matter body, are perfectly compatible with current physics and other branches of science. This is so, because Shadow Matter belongs to the world of physics and the discourse of science (see section 1.1). Since there exists, as yet, no definitive physics of Shadow Matter, I shall use the notion of Shadow Matter purely formally. I shall suggest that with the help of a few particular assumptions made about the physical properties of Shadow Matter a host of phenomena and case histories of parapsychology could be given straight forward explanations. Conversely, my theory suggests what sort of properties Shadow Matter could be expected to have if it is to explain psi-phenomena, hypnotic states and some aspects of the physical basis of normal mental processes.

Apart from its general assumptions, the theory, when applied to any particular case histories requires case-history-specific assumptions, so-called *ad hoc* hypotheses. This state of affairs, however, is typical of all scientific theories, notably theories such as Newtonian mechanics, classical electro-magnetic theory, quantum mechanics, quantum electro-dynamics and so forth.[7]

The present theory might put also metaphorical nooses round the necks of those prejudiced people, who believe that psi-phenomena are incompatible with physics, and, hence, impossible and cannot exist. Those who believe this, and others, imply also, almost invariably, that *all* claims for the existence of psi-phenomena are based alternatively on fraud, conjuring tricks, methodological errors, gullibility or plain stupidity (see James Randi 1982 & 1991). The new theory suggests that such dismissals of psi-phenomena and their observers may have been premature. As premature as the many prejudiced denials of the genuineness of hypnotic phenomena during a sizeable part of the nineteenth century. Randi may be a fascinating popular entertainer, but I am not sure that his skills are sufficient to judge the nature of complex phenomena that call for scientific explanations.

In 1970 Beloff proclaimed boldly 'that we must abandon hope of a physical explanation of psi-phenomena.'[8] This view may come naturally to people who, like Beloff, believe that human beings consist of a material body and an immaterial soul, which interact.[9] Such people also believe that the immaterial soul is needed to explain psi-phenomena (See also my remarks on a material 'soul' and psi-phenomena in the Preface). Unlike Shadow Matter, the immaterial soul does not belong to physics and has no physical properties. Dr Beloff, a former president of the Society for Psychical Research (London), believes, apparently, that we know already all possible answers that physics can supply – it is a view that I do not share.

I must stress also that if, as I maintain, psychic phenomena provide the best, and perhaps only, observable evidence for Shadow Matter, then this has important consequences. According to Kolb *et al* (1985) Shadow Matter is linked closely with a very particular string theory. Namely, it is related to that string theory which has gauge group $E_8 \times E_{8'}$. The reader does not have to worry what precisely is meant by a 'gauge group'. It is a very complex mathematical concept (see also my remarks in section 1.1) which is not required for understanding the present theory. Yet, because Shadow Matter is a possible consequence of a particular string theory, it follows that this string theory merits further extensive study.[10] This type of superstring theory might, conceivably, put us on the way to the Holy Grail of physics: *A Theory of Everything*. If I am right, then psi-phenomena, by bringing out the importance of Shadow Matter, may indirectly help to bring us closer to this mighty goal.

It is strange that I was driven to invoke Shadow Matter by my theorizing about telepathy and clairvoyance.[11] It was only after I published that paper that I began to study 'out of the body experiences' (OBEs). I had previously never attached much importance to OBEs. But then, suddenly, I realized OBEs also seem to suggest, as one possibility, that there could exist material systems, in addition to our ordinary matter bodies, which are detachable from these bodies. I recognized soon that Shadow Matter, as I had introduced it in my earlier paper,[12] could provide the additional material systems that become detached from ordinary matter bodies during OBEs. This, in turn, led me to suggest that our intellectual functions, our memories, our feelings and our perceptions could all be properties of the structure and functioning of the Shadow Matter brain - not of the ordinary matter brain, as has been believed for many hundreds of years (possibly thousands of years). But then people believed also for thousands of years that the Sun moves round the Earth. The time-span of existence of a belief is no guarantor of its correctness.

As a consequence of the present new view there arises now the possibility that after death and decay or destruction of the ordinary matter body the Shadow Matter body and its Shadow Matter brain could live on.[13] The surviving Shadow Matter brain, which, by assumption, is the carrier of our intellectual faculties, could preserve our existing memories, and have the capacity to form new memories. It could permit us to continue to think, judge, feel and (up to a point) perceive. It is no good arguing that such a thing could not happen. It happens, probably, albeit for vastly shorter time spans than even temporary survival would involve, during OBEs. The new theory, as will be shown in chapter 5, demolishes some of the most famous arguments against the possibility of survival of the human personality after bodily death.

I must at this stage add a short aside for physicists and other related scientists. Even general readers may ask, why should after death the ordinary matter brain decay, while the Shadow Matter brain could continue to exist stably, and possibly indefinitely? It is known to physicists that for an isolated ordinary matter system the Second Law of Thermodynamics proclaims that the system's state of disorder can either remain the same or increase. If this were also to apply to the Shadow Matter brain then one could expect its decay after death. Yet, there are no *a priori* reasons why the Second Law of Thermodynamics should apply to Shadow Matter. Hence, for all we know the Shadow Matter brain could persist stably after death of the ordinary matter body.

The present theory envisages that the Shadow Matter body and its Shadow Matter brain form more stable ensembles than their ordinary matter counterparts. Even if an ordinary matter body is cremated, its corresponding Shadow Matter body could remain intact. Such a forced separation of ordinary matter body and Shadow Matter body would, of course, strip the atoms of the ordinary matter body off their previously bonded Shadow Matter atoms (satoms). But, according to the present theory, this would also happen in an 'out of the body experience.' In an OBE this is a transitory state, and the atoms can reunite with their corresponding satoms, in an orderly manner (see section 2.1.1), when the Shadow Matter body reunites with the ordinary matter body, as the out of the body state comes to an end. If this view is correct, then Shadow Matter bodies and Shadow Matter brains in particular, would not only constitute models of corresponding ordinary matter bodies (and their brains). But more generally Shadow Matter objects could form models of all kinds of ordinary matter objects in the universe and could survive destruction of these ordinary matter objects.

In an expanding universe, like ours, one could expect, however, that there occurs in time an increasing attenuation of the density of free Shadow Matter throughout the universe. This state of affairs would not change even if there existed a reconversion of Shadow Matter objects into free Shadow Matter.

That Shadow Matter objects, such as Shadow Matter bodies of human beings, could persist independently of the ordinary matter objects that originally lead to their genesis is suggested by 'out of the body experiences' (see above). Thus the possibility of survival of the Shadow Matter body after ordinary bodily death seems reasonable. What seems more questionable is the duration of such survival. In the ordinary matter world of organisms there seem to have evolved aging processes which ensure that the organisms of any particular species have on average a more or less definite life span. Such mechanisms may have evolved to confer a great evolutionary benefit on various organisms. In the absence of such mechanisms the Earth surface could have become overpopulated with organisms, in the competition for food, quite apart from the struggle for existence. Whether comparable mechanisms of senescence have evolved in the Shadow Matter world remains, of course unknown. But some such mechanisms, mainly those concerned with a particular turnover of particular Shadow Matter structures, seems likely, as I shall argue in section 2.2.1. Whether this turnover applies also to Shadow Matter objects such as Shadow Matter bodies of human beings, thus converting these Shadow Matter objects gradually back into free Shadow Matter, remains uncertain. If, as I suggested, Shadow Matter bodies of human beings can exist almost autonomously during an 'out of the body experience' and, thus, survive autonomously, then turnover of Shadow Matter bodies cannot be a very rapid process, if it occurs at all. The only 'food,' if you wish to call it that, for which Shadow Matter objects compete during their genesis is free Shadow Matter, according to this theory. Likewise, there is no obvious need for competition for living space among possible surviving Shadow Matter objects. If there should exist a quasi-metabolic cycle in the Shadow Matter world then this could be as follows. Free Shadow Matter has combined with ordinary matter objects (including ordinary matter bodies) to form Shadow Matter objects. After destruction of the ordinary matter objects (including ordinary matter bodies) their previously attached Shadow Matter objects become detached and could survive for a while. After this they could be gradually converted, with the help of appropriate Shadow Matter machinery, into free Shadow Matter. The latter view would only be needed if the second law of thermodynamics applies also to Shadow Matter. This, however, as stressed above, need not be the case. Thus, while survival seems possible, its duration could be indefinite, or a matter of, say, years or decades or centuries or longer.

If Shadow Matter brains could survive indefinitely then this could provide one great evolutionary advantage for the mode of genesis of these brains. It could ensure that the 'best' surviving Shadow Matter brains of a great many successive generations could collaborate telepathically, thus forming 'superintelligences' and solve ever increasingly complex problems in mathematics and the sciences, and inspire great new works of art, and so forth. In fact, I suspect that some forms of inspiration, and amazing intuitions, could be due to occasional telepathy between receptive living people and such surviving superintelligences.

Although G. N. M. Tyrrell[14] did not suggest that inspiration might often originate from surviving superintelligences, and be telepathically communicated by them to living people of genius, he cites many relevant passages which are consistent with this thesis. He quotes mainly from Dr Rosamund E. M. Harding's book *An Anatomy of Inspiration*.[15] Thus:

Wordsworth told Bonamy Price that the line in his ode beginning 'Fallings from us, vanishings,' which has since puzzled so many readers, refers to those trance-like states to which he was at one time subject. During these moments the world around him seemed unreal and the poet had occasionally to use his strength against an object such as a gatepost to reassure himself. . . 'William tired himself with hammering at a passage,' wrote Dorothy Wordsworth.

I have discussed in section 1.6 below how trance-like states could arise and be connected to telepathy and/or clairvoyance. The possibility, therefore, that Wordsworth's inspiration, which seemed repeatedly related to trance-like states could have been triggered by telepathic information from surviving superintelligences, seems to fit my thesis. Again Tyrrell notes that:

Dickens declared that when he sat down to his book, 'Some beneficent power showed it all to him.' And Thackeray says in the *Roundabout Papers* 'I have been surprised at the observations made by some of my characters. It seems as if an occult power was moving the pen.'

Wagner discovered the opening of *Rheingold* during half-sleep on a couch in a hotel in Spezia; and in a letter to Frau Wesendonck he refers to the blissful dream-state into which he falls when composing.

Here, once more, we have an altered state of consciousness to be compared with Wordsworth's trance-like states when writing some of his works. Again, from Harding (1942):

George Sand, after describing Chopin's creation as miraculous and coming on his piano suddenly complete or singing in his head during a walk, says that afterwards 'began the most heart rending labour I ever saw. It was a series of efforts, of irresolutions, and of frettings to seize again certain details of a theme he had heard,' he would 'shut himself up in his room for whole days, weeping, walking, breaking his pens, repeating and altering a bar a hundred times' and spending six weeks over a single page to write it at last as he had noted it down at the very first.

George Eliot told J. W. Cross that in all that she considered her best writing, there was a 'not herself' which took possession of her, and that she felt her own personality to be merely the instrument through which this spirit, as it were, was acting.

If there exist, indeed, surviving Shadow Matter brains, and if these resemble, in performance capacities, the assumed detached Shadow Matter brains in 'out of the body experiences,' then we could expect that the surviving Shadow Matter brains produce the increased clarity of memory recall and the striking sense of reality which is typical of many 'out of the body experiences' (see section 1.6). The surviving Shadow Matter brain, liberated from its bonds with the ordinary matter brain, may reach greater heights of intellectual performance than during its bound existence. And if several such *liberated* Shadow Matter brains collaborate telepathically this could, indeed, become a source of inspiration.

As I expressed already in the Preface, the new world of Shadow Matter could cast entirely fresh light on the realm of the occult, and thereby provide a physical foundation for some parts of the occult. The Shadow Matter body could take the place of other *bodies* hypothesized by occultists. There is no need to have a multiplicity of kinds of such bodies; one variety, namely Shadow Matter bodies, could suffice. We may conclude that the two (or more) entity worlds of various religions (e.g. the dichotomy of body and soul) and of occultists may have contained an important grain of truth, but that the new scientific world of Shadow Matter may lead far beyond this.

Notes to Introduction

1. W Macneile Dixon, *The Human Situation* 1937, p. 429
2. A popular science book titled *Superstrings: A Theory of Everything?* edited by P C W Davies and J Brown, (Cambridge University Press 1988), is based on a BBC Radio 3 broadcast early in 1988 of interviews with distinguished theoretical physicists involved in the development of string theory. The book discusses also Shadow Matter.
3. Some people have argued that what people experience in OBEs are hallucinations. I shall argue repeatedly that to attribute OBEs to hallucinations seems highly implausible, when they are examined in detail. See pages 14, 27-29, 33, 37-8, 42, 44, 47-49.
4. In fact, in an earlier book (Wassermann, 1978) I championed still firmly the conventional point of view, although I realized that my earlier theory could not explain psi-phenomena.
5. Goddard 1986; Goelet *et al* 1986
6. See Penrose, 1976a, 1976b
7. See Wassermann, 1989
8. Beloff, 1970
9. Beloff, 1978
10. See Green and Schwarz (1984) for drawing earlier attention to this particular string theory
11. Wassermann, 1988
12. Wassermann, 1988
13. See page 16 and Wassermann 1988
14. 1948 Chapter 2
15. 1942, p. 68

1
Shadow
Matter
as the
Source
of
Psychic Phenomena

1.1 From Strings to Shadow Matter

Already in the Introduction I pointed out, briefly, that superstring theory led to Shadow Matter theory. Since Shadow Matter plays such a central part in this book, let me say just a little more about how Shadow Matter originated from string theory. For centuries, since Sir Isaac Newton's days, and perhaps earlier, people have introduced 'point particles' into physics. A point particle is a piece of matter concentrated in one point. Later, when 'elementary particles,' such as the electron, were discovered they were assumed to be point particles. For many purposes this assumption seemed adequate. Yet, to consider elementary particles as point particles led to some technical difficulties in some parts of theoretical physics. This inspired some people, notably M. B. Green (see 1985), to investigate 'strings'. String theory assumes that elementary particles are not point-like, as used to be thought. Instead string theory assumes that each elementary particle is composed of an extended minute string-like entity (see Green, 1985; Ellis, 1987). The mathematical theory of such strings is extremely complex (see Green and Schwarz, 1984; Green *et al.* 1987). String theory tries to predict (or explain) all the basic properties and possible modes of interactions of the known, and perhaps still to be discovered, elementary particles. This should lead to a unified theory of physical forces and matter often called a *Theory of Everything* (or TOE for short). To construct a TOE is the goal of some contemporary String physicists (see Green, 1985; Green *et al.* 1987; Ellis, 1986, 1987).

This is almost all that the reader has to know about string theory, except for a few very important points. By 1984 all, except two, of the then known string theories showed serious mathematical anomalies (Green and Schwarz 1984). Among the anomaly-free string theories was one, which became known technically as having 'gauge group' $E_8 \times E_8$. The notion of a 'gauge group' requires very advanced mathematical knowledge and need not concern us here (see Green *et al.* 1987; Green, 1985). The gauge group $E_8 \times E_8$ led, on purely mathematical grounds, to the possibility that, in addition to ordinary matter of the familiar kind, of which we, our houses, tables and chairs are made, there could exist another kind of matter, called Shadow Matter by Kolb *et al.* (1985).

So it was exactly one type of string theory that led to Shadow Matter. This tells us something about the roots of Shadow Matter theory and its relation to string theory. Ordinary matter is believed to be constituted ultimately of elementary particles of a few kinds (e.g. quarks and electrons (see p. 12). Similarly, Shadow Matter could be built up from elementary Shadow Matter particles (see p. 12)). Elementary Shadow Matter particles could interact with each other, apart from gravity, by new types of forces which could differ from most of the types of forces familiar from ordinary matter physics.

Kolb *et al.* (1985) considered several possible versions of Shadow Matter theory, all based on *ad hoc* assumptions, notably about the weights of their assumed elementary Shadow Matter particles. While Kolb *et al* (1985) guessed probably correctly, on the basis of string theory arguments (see above), that there exists Shadow Matter, this does not tie us down to their somewhat arbitrary assumptions about Shadow Matter. I shall accept, therefore, the existence of Shadow Matter as a good bet. Also, throughout the remainder of this book, I shall argue that Shadow Matter could explain most or all psychic phenomena, provided Shadow Matter has various assumed properties, which are my assumptions. [My theory, therefore is almost entirely independent of the theory of Kolb *et al*, and rests largely on different assumptions, partly suggested by various properties of psychic phenomena. Because of this independent standing of my theory, I am not obliged to derive my theory from, or relate it to, the theory of Kolb *et al*. The points just made merit attention. An American philosopher, who believes that psi-phenomena do not exist, believed, equally mistakenly, that the assumptions of Kolb *et al* are necessary, and that, therefore, I ought to base my theory on their assumptions and derive my theory from their theory. This demand, of course, is completely wrong. All that my theory and that of Kolb *et al* (1985) have in common is the shared belief that Shadow Matter exists, as a result of string theory. Also, like strings, Shadow Matter belongs to the realm of mechanistic materialism (see p. 12 for definition).]

Let me now introduce some assumptions concerning Shadow Matter, assumptions mentioned already briefly in the Introduction. It is well known from physics and chemistry that ordinary matter can interact with other ordinary matter in a variety of ways. In suitable conditions, specific ordinary matter can get bound to other ordinary matter. Typically paint can get bound to a canvas. Some of the best known general examples of the bonding of ordinary matter to ordinary matter occur, on a wide scale, in chemistry, in the form of atomic and molecular bonding. Chemists explain, by their theories, how specific atoms of one kind, say hydrogen atoms, can get bound to specific atoms of the same kinds, or other kinds, say carbon atoms, so as to form particular molecules. I assume here that particular sub-structures of Shadow Matter can become bonded also to one or more kinds of other particular substructures of Shadow Matter. I assume also, as Kolb *et al* (1985) *did not do*, that Shadow Matter can get bonded gravitationally to ordinary matter and that this has wide implications for my theory of psi-phenomena. In fact, the selective bonding of Shadow Matter to ordinary matter, together with the emission of Shadow Matter from complexes of bonded ordinary matter and Shadow Matter, could, together with various subsidiary hypotheses,[1] account for all major psychic phenomena, as I shall demonstrate below.

I have discussed already how Shadow Matter might get bound to ordinary matter (p. 12). Thus, the possibility that all ordinary matter, in nature, can become bound to, and often combine stably with 'ghostly' Shadow Matter becomes an important physical possibility. The notion that 'we each possess intermingled with our physical [ordinary matter] organism a duplicate subtle body of some exceedingly tenuous matter (see Mead, 1919), a body which may be or become the vehicle of our mental attributes and which may survive dissolution of our fleshly [ordinary matter] bodies (cited from Gauld, 1977, p. 601) is by no means new (see p. 20 and Preface). What is new here, is the identification of the 'tenuous matter' with Shadow Matter (see p. 22), which, in turn, may be part of an ultimate *Theory of Everything* (see p. 21). Equally new is the much wider assumption that Shadow Matter can not only bind to human bodies, but that appropriately matching Shadow Matter can become bound, in close proximity, to all kinds of ordinary matter, animate and inanimate (see pp. 12-3). In fact as in Wassermann (1988), I suggest that psi-phenomena are not confined to living creatures. By assumption these phenomena are a universal property of all appropriately matching Shadow Matter, whether bound to living ordinary matter or to inanimate ordinary matter (see p. 12).

This point of view differs drastically from that of the late, prominent, parapsychologist J. B. Rhine. He asserted (Rhine, 1977, p. 169) that 'only living creatures manifest psi-capacity.' In fact, if inanimate ordinary matter could not, say via Shadow Matter bonded to it, emit 'psi-signals,' then it would be hard to grasp how living organisms, say people, could become normally aware clairvoyantly (i.e. without use of normal sensory cues and without the intervention of other people's Shadow Matter brains via telepathy) of states of inanimate objects often many miles away from them. Indeed, if all ordinary matter could bind matching Shadow Matter, then it becomes tempting to assume that 'psi-signals' are produced indirectly by the Shadow Matter that is bound to ordinary matter. How this could occur will be described later. Much of this book aims to explain in great detail how Shadow Matter could produce psychic phenomena of the most diverse kinds.

1.2 Some Spontaneous Psi-Phenomena and their theoretical Implications.

Continuing wide interest in psychic phenomena is not, as sceptics often claim, an indicator of far-ranging superstition, irrationality, public gullibility or general stupidity. On the contrary, irrationality seems to rest, at least in my experience, with those who most vehemently reject the genuineness of *any* psi-phenomena. Probably many normal people have had some psychic experience. But they are often ashamed to admit that they believe in the occurrence of such phenomena. They fear that they might become the laughing stock of others, who attribute such phenomena to 'mere chance coincidences' or to 'hallucinations'. Likewise, acceptance of the genuineness of hypnotic phenomena was regarded as something disreputable early last century and even later (see section 1.11). Yet, for no particularly obvious reason, people gradually accepted the occurrence of hypnotic states as genuine, possibly because the medical profession said so. Yet, the biological nature of hypnotic states and the physiology and/ or physics underlying them was hitherto not much better understood than a century ago. Thus, publically professed attitudes, including taboos, can change like fashions, depending on the academic and public climate. As long as some professors of psychology indoctrinate a whole lot of students with their prejudices against some particular subject, and these students do not study the subject first hand, considerable damage can result to the progress of such a subject. I believe, rightly or wrongly, that there are some academic psychologists who regard themselves as guardians of 'normal psychology' and who consider it their duty to assail parapsychology whenever possible.

Some spontaneous psychic phenomena seem to turn up less frequently than others. Rarity of phenomena is apt to contribute to scepticism concerning their occurrence. Although lightning, in any particular locality, may be also a relatively rare event, yet, when it occurs, it can often be witnessed simultaneously by many people. Accordingly, lightning, although not repeatable on demand, is accepted as a natural phenomenon. Unfortunately, this does not apply to many psi-phenomena. Like people's unrecorded thoughts, these phenomena can often only be witnessed by a single percipient. Notable exceptions are cases of accurate information provided by certain 'psychic mediums' (often during 'trance states') in public, in cases where the medium could not have obtained the information by ordinary means. Other exceptions are collectively perceived apparitions, where, to cite Gauld (1977, p. 602): 'Two or more persons simultaneously see what is *prima facie* the same phantasmal figure in the same place' (see Hart and Hart (1933) for a review of such cases).

Let me just cite here one such case, taken from the *Journal of the Society for Psychical Research* (London) vol. 6, p. 129 and summarized by Tyrrell (1948, p. 63) as follows:

Case 1

A certain Canon Bourne and his two daughters were out hunting, and the daughters decided to return home with the coachman while the father went on. 'As we were turning to go home' say the two Miss Bournes in a joint account, 'we distinctly saw my father waving his hat to us and signing us to follow him. He was on the side of a small hill, and there was a dip between him and us. My sister, the coachman and myself all recognized my father and also the horse. The horse looked so dirty and shaken that the coachman remarked he thought there had been a nasty accident. As my father waved his hat I clearly saw the Lincoln and Bennett mark inside, though from the distance we were apart it ought to have been utterly impossible for me to have seen it. . . Fearing an accident, we hurried down the hill. From the nature of the ground we had lost sight of my father, but it took us very few seconds to reach the place where we had seen him. When we got there, there was no sign of him anywhere, nor could we see any one in sight at all. We rode about some time looking for him but could not see or hear anything of him. We all reached home within a quarter of an hour of each other. My father then told us that he had never been in the field, nor near the field in which we thought we saw him, the whole day. He had never waved to us and had met with no accident. My father was riding the only white horse that was out that day.'

Tyrrell, I think rightly, considers this as a telepathically produced collective apparition, although he states that 'the cause which set the telepathic machinery in motion in this case is obscure.' According to my type of theory, to be developed more fully in chapter two, telepathic information, whether, as in the case just cited, in the form of a minutely detailed full hallucination of people and scenery or in the form of vaguer impressions, could be generated by the Shadow Matter brain. This could happen when the Shadow Matter brain is acted on by externally arriving Shadow Matter signals generated by other Shadow Matter brains. One could argue that, in **Case 1** above, the basic hallucination was generated by the Shadow Matter brain of one of the three percipients and was then by 'telepathic machinery' communicated to the Shadow Matter brains of the other percipients. How such 'telepathic machinery' could operate by means of Shadow Matter will be discussed in chapter two. What matters at this stage is that if particular Shadow Matter systems are the exclusive physical representatives, carriers, senders and receivers of telepathic messages, then the Shadow Matter brains and Shadow Matter signalling systems involved must be capable of the same degree of accurate representation of minutiae of scenery, people, animals etc, as occurs in ordinary visual perception. This is consistent with one of my main assumptions. I assume that the Shadow Matter brain can function in much the same way in normal perception, when the ordinary matter brain acts on it, as when Shadow Matter psi-signals induce hallucinations in that Shadow Matter brain. I assumed in the Introduction (p. 15) that in 'out of the body experiences' the Shadow Matter brain becomes (partly or completely) detached from the ordinary matter brain and body and can then, via Shadow Matter eyes, or alternatively directly, receive Shadow Matter signals.

In **Case 1**, in order to recognize the rider as Canon Bourne, or the white horse as a white horse, or to recognize the identity of the field in which Canon Bourne's apparition was seen, the following could be required. The hallucinatory machinery of the percipient must have had access to the percipient's memory traces (i.e. stored memories) of the Canon's outward appearance. This seems necessary in order to match the physical (Shadow Matter) representation of the Canon's apparition with the memory traces. Indeed, this seems required by the telepathic receiver in order to identify the apparition with a normal memory-stored image of what the Canon looked like (and similarly for the other things recognized in the apparition). In other words, telepathically induced apparitions, in order to be meaningful, must, like ordinary things that are viewed and recognized, have access to memory traces. Thus, the same memory traces that are available to normal perceptual machinery must also be accessible to psi-perceptual machinery. It is such considerations that led me to the suggestion (see p. 14) that psi-perceptual machinery and normal perceptual machinery share certain components. This suggestion, in turn, led me to assume that normal perception and hallucinations in psychic phenomena (including when they occur in 'out of the body experiences' (see **Case 5**), are *all* produced by the Shadow Matter brain.

Although **Case 1**, or other reported collective apparitions, provide indirect evidence for telepathy, these cases, while often impressive, are not as numerous as cases of other types of psychic phenomena. Quite often, perhaps in most cases, psi-phenomena may occur 'wrapped up' in the guise of 'synchronicity experiences' (SYNEXs) more fully discussed in chapter four. Let me give here just one example of a SYNEX. This SYNEX could have a perfectly normal (non-psi) explanation and is not as impressive as many SYNEXs cited later. It comes from my own collection of SYNEXs.

Case 2

On the 30.1.1988 I was thinking of Mrs R. (the sister of a friend) who lived then in Adelaide Road near Chalk Farm, London, and whose address I had found out only recently. While I thought of Mrs R., I thought also that it seemed, as far as I could remember, that I had only heard of Adelaide Road apparently for the first time recently, namely in connection with Mrs R. On the following day, 31.1.1988, I found an old issue of the *Proceedings of the Society for Psychical Research* vol.50, May 1953, on a table at home, where I had left it some weeks earlier in order to get some totally different information. I decided to re-read an article by S. G. Soal in that issue, titled 'My thirty Years of Psychical Research,' having not read the article for many years. On page seventy-five Soal writes, in connection with some extrasensory perception experiments, that 'During the first year (October 1927-July 1928) a small group of five or six agents or senders, which included myself [ie. Soal] met regularly for half an hour on the same evening of each week originally at the SPR rooms and later on, at the house of Professor and Mrs J Mackenzie at No. 2 *Adelaide Road*, Chalk Farm.' (Italics are mine.)

Here, as is typical of other SYNEXs, which others and I have collected, one or more specific, often unusual, symbols (in this case Adelaide Road and Chalk Farm, which I had thought of also on 30.1.1988) surfaced for the same percipient twice within two days or less, and in entirely different contexts.

It could be argued that Soal's article, when I first read it many years earlier, left memory traces of Adelaide Road and Chalk Farm. These traces could have caused me to re-read Soal's article shortly after my memory was stimulated by the symbols Adelaide Road and Chalk Farm, which I had thought of one day earlier. Yet, the two symbols, as such, are of little importance in relation to the major subject matter discussed by Soal. Hence, unless it is argued that normal people, including myself, remember every line of print they have ever read (i.e. store it unconsciously as memory traces), then it would seem very unlikely that I had formed earlier an associated memory trace of the two printed sets of symbols. In fact, there exists no evidence from the psychology of memory that ordinary people (who are not hypnotized) have such 'super memories'. Moreover, other SYNEXs, cited in chapter 4, could not be explained in this way. Alternatively, I could have become aware clairvoyantly of the symbols Adelaide Road and Chalk Farm in the volume on my table, and, because they matched the corresponding symbols consciously experienced on the previous day, I looked up the article by Soal. On the face of it this may seem less likely to some readers than an explanation based on a perfect memory system for anything ever seen in print. Yet, examples of other SYNEXs suggest that such an astonishing degree of clairvoyance may have operated in **Case 2** and in many other SYNEXs (see chapter 4).

One could, of course, attribute all SYNEXs to mere chance coincidences. More generally one could urge that all evidence for telepathy and clairvoyance is due to chance coincidence, and that cases of 'out of the body experiences' are mere 'hallucinations'. There is, however, the fact that most or all psychic phenomena fall into classes with repeatable class characteristics. This suggests that psi-phenomena, although spontaneous in the cases considered here, may be caused by mechanisms (Shadow Matter mechanisms to be precise) which produce on different occasions phenomena with similar class characteristics. Hence, some people might prefer causal explanations of psi-phenomena as alternatives to chance coincidence arguments, provided these explanations can be given within a coherent physicalistic theory. The physicalistic theory of psi-phenomena presented in this book, of course, does not only deal with SYNEXs but explains a host of other kinds of evidence for psi-phenomena (see Case 1).

Judging by my own observations and those of the psychotherapist C. G. Jung (1955) (whose interpretations of SYNEXs I do not accept, see chapter 4), SYNEXs of high structural content are of relatively common occurrence among human populations. Contrary to Jung, SYNEXs could provide excellent evidence for psi-phenomena. I have experienced some SYNEXs following each other within hours, while I experienced the next SYNEX only several weeks or months later. On the whole SYNEXs seem to occur clustered, with several coming over a span of several days, followed by a pause, which may be weeks or months.

According to section 1.6 my theory claims that weak bonding of the Shadow Matter brain to the ordinary matter brain favours the occurrence of psi-activation of the Shadow Matter brain. By contrast, strong binding of the Shadow Matter brain to the ordinary matter brain, which normally is assumed to be the rule, strongly inhibits the occurrence of psi-phenomena. This suggests that the long absences of SYNEXs correspond to period of (normal) strong binding of the Shadow Matter brain to the ordinary matter brain. The short periods of multiple occurrences of SYNEXs could correspond to short periods of temporary weakening of some or many of the bonds between Shadow Matter brain and ordinary matter brain. *If SYNEXs were simply due to chance coincidences, one would not expect such a clustering, followed by spacing, of SYNEXs.*

Most psi-phenomena are experienced by single individuals, or, in some cases (see above) occasionally, as in **Case 1**, by a few individuals. This does not mean that we cannot study and classify psi-phenomena. Many of the most clearly noticeable psi-phenomena occur spontaneously and cannot be produced to order or be rigidly controlled. Their spontaneity is more like that of lightning, earthquakes, volcanic eruptions or rare cosmic ray events. Nevertheless, the fact that one can classify these psi-phenomena, ie. that they seem to fall into a number of clearly discernible classes, suggests that their typical class characteristics, and apparent appropriate repeatability of class-specific features, are not due to chance, but involve underlying mechanisms. For instance, SYNEXs seem to be typical as regards the resurfacing of the same meaningful, often unusual, symbols, within a short time (up to two days or often much sooner). Yet, for those who feel inclined to attribute to chance everything they cannot explain, there are, apart from SYNEXs and apparitions, a host of other kinds of classifiable phenomena that seem either to implicate psi-mechanisms, or, at least, could be satisfactorily explained in terms of such mechanisms.

'Out of the body experiences' (called OBEs for short) are prominent among spontaneous psi-phenomena with repeatable class characteristics (see Celia Green [1968] for many excellent case histories). OBEs occur often during near death states. According to a recent BBC broadcast (1.2.1988) on Radio 4, a Gallup poll indicates that many thousands of people in the USA have had a OBEs at some time in their lives. These OBEs were, again, 'single person experiences,' and many involved near death of people having major surgery or accidents that caused them to be unconscious. The striking resemblance of the class characteristics of these massive numbers of cases makes OBEs as credible as lightning that is witnessed by a sizable population of a town. There are, of course, people who argue that OBEs are simply hallucinations. This, however, does not explain why the OBEs of so many different of people have apparently most or all the same repeatable class characteristics (see Moody 1976 and Sabom 1982 and also pp. 37 ff for listing of these characteristics). Are we to assume that a large number of people hallucinated all some or many of the class characteristic features of OBEs? Most typically percipients of OBEs seem to see their own body from the outside, as if they did not form part of that body any more. Why should such a large number of people, independently of each other, 'hallucinate' such an external viewing of their own body? Similar questions can be raised about other aspects of many OBEs.

Sabom at first seemed sceptical of Moody's findings, but confirmed most of the class characteristic features of near death state OBEs in his own observations, while attached as a medical specialist to an intensive care unit of a hospital. I shall now, and in the following sections, cite a few case descriptions of OBEs, taken from various sources, some of the cases being separated from each other by the better part of a century. I shall use these case histories in order to suggest various new theoretical conclusions.

Case 3

The following typical OBE case is one of many given by Moody (1976 p. 34), from which I shall draw my own conclusions, which will be important in relation to my theory of OBEs and to my general theory of psi-phenomena. Moody wrote:

A man's heart stopped beating following a fall in which his body was badly mangled, and he *recalls* (italics are mine)

'At one time - now, I know I was lying on the bed there - but I could actually see the bed and the doctor working on me. I couldn't understand it, but looked at my own body lying there on the bed, and I felt real bad when I looked at my body and saw how badly it was messed up.'

Here, as in many other OBEs, the percipient of the OBE can not only perceive his own ordinary matter body from a position apparently outside that body and some distance from it. Percipients in this and some other cases could also, when appropriate, see themselves being manipulated, again, as if they were outside spectators (who in one typical case, for example, felt that she was located at the ceiling of her hospital room, (see **Case 4**)).[2]

Again, during the OBE percipients can have emotions such as 'feeling real bad' when seeing their own body badly mangled (in **Case 3**). Also, during an OBE percipients can make judgments about their situations and about other things, recognize in precise detail some of the personalities that handle their ordinary matter body, see things in natural colours, and remember also their judgments made during the OBEs, remember the people encountered and, occasionally, other places visited during their OBE, say within a hospital (see Green 1976 p. 113), while they were ostensibly 'outside' their ordinary matter body. In quite a few such cases the OBE percipients were, as regards outside judges, apparently unconscious, i.e. their ordinary matter brain was in a subnormal state. (Possibly the abnormal state of the ordinary matter brain during an OBE may have been instrumental in allowing the Shadow Matter brain to have become detached from the ordinary matter brain, and the rest of the Shadow Matter body could then have followed suit.)

Perhaps even more startling than the sighting of their ordinary matter bodies are reports, by several OBE percipients, that during their OBE they seemed to have a 'quasi-material body' of their own. With the help of this quasi-material body they could 'see' their ordinary matter body and move about (see **Case 5**). This quasi-material body was described as, e.g. jelly-like (see **Case 5**, see also Moody 1976, pp. 36-43). If such a quasi-material body did not exist and were purely hallucinatory, then one would have to explain why several different percipients (e.g. **Case 5** and Moody 1976) should independently of each other and, in one case, nearly a century earlier, have given (and hallucinated) similar descriptions of that quasi-material body. There are no reasons for believing that some of Moody's cases, who reported such a quasi-material body had looked up the relatively inaccessible case report of **Case 5**, as a basis of similar claims.

There are, however, some notable exceptions. First I note that in some cases of OBEs only a part of the quasi-material body may leave the ordinary matter body (see the interpretation of **Case 6**). For instance, the head might leave, leaving the rest of the quasi-material body behind. This interpretation is suggested by the following case history cited by Celia Green (1976 p. 117):

> I was in bed, it was summer, and I couldn't get to sleep... After a time I decided to turn over and try to sleep, in doing so I felt a quick movement inside me, I seemed to be inside myself leaping up to get out. Next thing I know is that I'm near the ceiling in the corner of the room looking down on my body with my husband next to it. At first I thought it very funny it couldn't be possible. I'm up here and yet I'm lying in bed down there. I wondered if I could wake my husband up but I seemed to have no hands to shake him or touch him, there was nothing of me, all I could do is see. Then the thought struck me that I might

be dead and I panicked. Immediately thinking that thought I slipped straight back into my body as quickly as I had come out.

As mentioned in the Introduction, my theory assumes that associated with the ordinary matter body is a Shadow Matter body, endowed, among much else, with a Shadow Matter head with Shadow Matter eyes, a Shadow Matter brain and the Shadow Matter equivalent of a nervous system. So if, in the case just cited only the lady's Shadow Matter head left her ordinary matter head, then her Shadow Matter eyes and Shadow Matter brain migrated together with the Shadow Matter head. Her Shadow Matter brain, while out of the ordinary matter body could still make judgments leading to 'amusement' and later 'panic'; and her Shadow Matter eyes could 'see' her ordinary matter body and that of her husband in bed. (The way in which Shadow Matter eyes could see with the help of Shadow Matter photons (=sphotons) is explained in section 1.5) If the rest of the lady's Shadow Matter body remained attached to her ordinary matter body, then she obviously had no free Shadow Matter arms.

I conclude that Celia Green's case, just cited and discussed, does not imply that we do not have a quasi-material body (which I shall repeatedly identify with the Shadow Matter body). It may mean simply that sometimes only a part of the quasi-material body leaves the ordinary matter body during an OBE.

1.3 Coherent Shadow Matter Models (SMMs) and 'Out of the Body Experiences.'

According to my previous assumptions (p. 13) the system of mutually linked Shadow Matter constituents, which are attached to an animate or inanimate ordinary matter body, could form a coherent Shadow Matter Model of that ordinary matter body. Such a coherent Shadow Matter Model will be called SMM for short. Those parts of a SMM that cohere could be held together by Shadow Matter forces.

In particular, according to this theory, a human ordinary matter body and all its organs and structures, including the minutiae of the brain, could give rise to an attached Shadow Matter Model (SMM) (p. 13). Normally, this SMM could remain tightly bonded to the ordinary matter human body, during most of its life-time. Such a SMM, bound to a human ordinary matter body will be called SMM(body) for short. That particular sub-SMM of SMM(body) which is normally attached to the ordinary matter brain will be referred to as SMM(brain). My theory implies that as a human being (or other organism) develops, its SMM(body) also develops. Addition of new ordinary matter components to the ordinary matter body, during growth and development, could add, automatically, corresponding Shadow Matter components which become bound to the added ordinary matter components. In turn, some of the added Shadow Matter components could become linked to the SMM(body) which exists already. When the ordinary matter body exchanges components with the environment (e.g. during metabolic turnover), SMM(body) components, bound to the turned over ordinary matter components could remain fixed to the SMM(body) and do not turn over. Thus, Shadow Matter memory traces of the SMM(brain) could be preserved during turnover of ordinary matter brain structure. (This interpretation differs drastically from that proposed by Crick (1984).)

My theory suggests (p. 13) that every kind of simple or complex ordinary matter system can be modelled and accompanied by a corresponding SMM(system). I am assuming, however, that any Shadow Matter constituent that models a particular ordinary matter constituent (say an electron), by becoming attached to that ordinary matter constituent has only a minute fraction of the weight of that ordinary matter constituent (see p. 13). Hence, a SMM(body) is almost negligibly 'light' compared to the ordinary matter body to which it is attached.[3] This, however, does not prevent, say, the SMM(body) from becoming normally tightly bonded to the ordinary matter of the human body via gravitational bonds (see p. 12). Although gravitational attraction between Shadow Matter and ordinary matter could be expected to be very weak, if it is directly proportional to the weight of the attracted Shadow Matter, the assumed minuteness of that weight could be compensated for by increased proximity between Shadow Matter and ordinary matter. This increased proximity could increase the gravitational attraction.

It could be that in an 'out of the body experience' (OBE) most, but not all of the SMM(body) becomes detached from the ordinary matter body, with the SMM(body) retaining its detailed coherent structure. The detached SMM(body) could then move off some distance away from the ordinary matter body into space, remaining only residually attached to the ordinary matter body by an elastic (and, hence, extensible) Shadow Matter 'cord,' which forms a bonding system between the detached SMM(body) and the ordinary matter body. The Shadow Matter 'cord' is believed to be essential for enabling re-entry of the SMM(body) into the ordinary matter body. Several OBE percipients have claimed that they 'saw' a cord (see **Cases 5, 7-7a**). Gauld (1977, p. 607-8) notes that the cord 'is mentioned as early as Plutarch's *De Sera Numinis Vindicta*.' The assumption that a SMM(body) becomes detached in an OBE is consistent with various observations.

First, the reported autonomous quasi-material body (p. 29) could correspond to the detached SMM(body). Moreover, the experienced movement of the quasi-material body away from the ordinary matter body during an OBE (see Moody (1976), Sabom (1982) and **Cases 5** and **7**) would be consistent with the assumption that it is not a purely hallucinated entity, but corresponds to the detachable SMM(body) which can move about. Second the subjective feeling that the experienced 'I' (or personal identity feeling) of the subject during an uncomplicated OBE (in which the personality does not split) is experienced as being linked to the quasi-material body (where this is reported) is significant. By contrast the 'I' which is normally experienced as linked to the ordinary matter body is present when there is no OBE. This is consistent with the assumption that SMM(brain)s and not ordinary matter brains form all (normal and OBE) conscious experiences as byproducts (so-called 'epiphenomena') of some of their processes. Indeed, if the subjective, normally experienced 'I' is always produced by the SMM(brain) rather than by the ordinary matter brain, then this leads to the following conclusion. The assumed detachment and, relatively short-distance, removal of the SMM(body) (including its SMM(brain)) from the ordinary matter body during an OBE would lead to the experienced 'I' moving away from the ordinary matter body while remaining associated with the SMM(brain).

There seems to be a good deal of circumstantial evidence which is consistent with the assumption that in some OBEs only part of the SMM(body) leaves the ordinary matter body (see my comments concerning the case cited from Celia Green (1976) on p. 29). In some cases this 'splitting' of the SMM(body) may lead to only part of the SMM(brain) leaving the ordinary matter brain (see section 1.8 for one such case). Such cases may manifest themselves in the appearance of 'multiple personalities'. I shall discuss briefly one such case at the end of this Section.

Third, last, but not least, is the fact that percipients of OBEs perceive not only their ordinary matter body from a position outside their body. They perceive also some of the minutest details of the environment of their ordinary matter body, in full colours, as if 'seen' from positions, apparently, occupied by their quasi-material body outside their own ordinary matter body. What they seem to experience via their quasi-material body (here identified with part or the whole of the SMM(body)) is just the sort of thing they could experience if their ordinary material body were (which it is not) located in the position of the experienced quasi-material body (in cases where this is experienced), and if their ordinary material body could, with perceptual machinery, see their environment from that position.

Let me now cite part of a case history of an OBE which suggests that in *some* OBEs there may occur a splitting of the assumed SMM(brain) into one part that remains attached to the ordinary matter brain and another part that leaves the ordinary matter body. The case comes from the *Journal of the Society for Psychical Research* (London) vol.25, p. 126 and is cited by Tyrrell (1948, p. 196). The case was copied by Mr. Norman F. Ellison from a diary which he kept during the First World War. Ellison and a companion 'H' had been subjected to enormous strain in reaching new trenches, under abysmal conditions. Ellison continues:

Several hours of this misery passed and then an amazing change came over me. I became conscious, acutely conscious that I was outside myself; that the real 'me' - the ego, spirit or what you like - was entirely separate and outside my fleshly body. I was looking in a wholly detached and impersonal way upon the discomforts of a khaki-clad body, which, whilst I realized that it was my own, might easily have belonged to someone else for all the direct connection I seemed to have with it. I knew that my body must be feeling acutely cold and miserable but I, my spirit part, felt nothing.

His companion told him that his grim silence had suddenly given place to wit and humour and he had chatted as unconcernedly as if before a comfortable fire.

This passage, cited from Tyrrell (1948), suggests, that, in terms of my theory, the part of Ellison's SMM(brain) which represents (in physical form) his 'I' (and epiphenomenally gives rise to his personal 'identity feeling') left the ordinary matter brain. By contrast that part of Ellison's SMM(brain) which controlled his speech and a part of his capacity for thought and feeling and memories (notably memories for the speech vocabulary) stayed behind with Ellison's ordinary matter brain.

The Ellison case is not an isolated case of its kind, and Tyrrell cites several similar OBE cases. Another of these will be discussed in section 1.8. Tyrrell, however, had no theory of psi-phenomena by means of which to explain such cases.

1.4 The Thesis that all Mental Functions are accomplished by the Shadow Matter Brain (i.e. SMM(brain))

There is now a choice. Either one can assume that, as just suggested, OBEs are produced by a detachable SMM(body) (including its SMM(brain)), or one could assume that there exists no SMM(body) and that the ordinary matter brain on its own hallucinates the OBE. I have argued already against the latter (section 1.2 and shall argue further in sections 1.6 and 1.7). Also, since in various OBEs the ordinary matter brain, on its own, could simply not have access to many things that are supposedly hallucinated during an OBE, one would have to assume that the ordinary brain has paranormal powers! It is for these reasons, and others mentioned above and below, that I prefer to assume the existence of a detachable SMM(body). Perhaps a decade earlier this assumption would have seemed totally implausible to most physicists and physiologists. With the discovery of string theory, however (section 1.1 p. 21), the possible and likely existence of Shadow Matter has become an attractive possibility. The postulate of SMMs may, therefore, now be more acceptable to physicists and physiologists than it would have been at earlier times. Moreover, as will be shown, the assumption that SMMs exist and participate in psi-phenomena, could explain a host of different psi-phenomena in addition to OBEs.

We come now to a central issue. Since SMM(brain)s are supposed to model most, possibly all, structural parts of ordinary matter brains accurately in terms of Shadow Matter, there exist the following possibilities. First, it could be assumed as by physiologists and many others that an ordinary matter brain, on its own, could provide, in terms of *ordinary* matter *hardware*, complete representations of memory traces and thoughts and feelings etc. Also in this traditional kind of theory the ordinary matter hardware, on its own, generates conscious experiences as byproducts (epiphenomena) of its activity. Shadow Matter theory could then assume that in addition to ordinary matter brains their attached SMM(brain)s could, while sufficiently tightly bonded to ordinary matter brains, and under the influence of ordinary matter brain machinery, form *secondary* Shadow Matter representations of all memory traces, thoughts, feelings etc., that the ordinary matter brain also represents in this kind of theory. Moreover, in this type of theory (used by Wassermann, 1988), while the SMM(brain) is tightly bound to the ordinary matter brain, the latter can exchange gravitons with the SMM(brain). Gravitons are minute 'quanta' (i.e. energy parcels) carried by 'gravitational fields' (see also p. 14). Since, by assumption (see Kolb *et al.* (1985) and p. 14) ordinary matter and Shadow Matter can only interact gravitationally, gravitons provide one important way of exchanging energy and information between an ordinary matter system and a Shadow Matter system. In the theory under consideration, when an ordinary matter brain and an SMM(brain) exchange gravitons, some of these exchanged gravitons could then assist in producing, within the SMM(brain), secondary representations of cognitive constructs (i.e. memories, thoughts etc.) which are also primarily represented within the ordinary matter brain.

Further, according to this possible first theory, when in an OBE a SMM(brain) becomes detached from the ordinary matter brain, then that SMM(brain) could continue to form cognitive constructs (e.g. memories) autonomously. Consciousness would then be a byproduct of SMM(brain) activity during detachment. I shall call this first possible theory the 'double representation theory.' This theory suffers from the weakness that it assumes an alternative, mutually exclusive, physical representation of mental states. While the SMM(brain) and the ordinary matter brain are tightly coupled, it is, according to the first theory, the ordinary matter brain that has conscious states. But in an OBE, according to this theory it is the SMM(brain) that forms memories, thoughts etc., and generates conscious states. Because of this, I prefer a second, alternative, theory.

This second theory assumes that memory traces and thoughts etc., are exclusively represented by the SMM(brain) in normal states *and* in OBEs. This theory assumes also that conscious experiences, when they occur, are always exclusively by-products of SMM(brain) activity and not of ordinary matter brain workings. It is assumed, however, that while the SMM(brain) is tightly bound to the ordinary matter brain, the SMM(brain) has restricted autonomy as regards formation of memory traces and other cognitive constructs. During such tight binding the ordinary matter brain, could, by means of emitted gravitons, induce formation of memory traces and physical representations of other cognitive structures by the SMM(brain). Thus, while the ordinary matter brain could determine what the SMM(brain) does in these circumstances, it is the SMM(brain) and not the ordinary matter brain that stores memory traces and represents other cognitive constructs. I shall call this second, alternative, theory the 'single representation theory,' since, according to this theory, thoughts etc., are only represented by one brain per person, namely by the SMM(brain) and not by the ordinary matter brain as well.

Also, by graviton exchange the SMM(brain) could induce synaptic changes of the ordinary matter brain that could lead to motor activities, e.g. speech. I shall return to this somewhat more technical point at a later stage. The theory assumes, however, that while the ordinary matter brain and the SMM(brain) are tightly coupled, activation and formation of memory traces and activation of the physical representatives of thoughts etc., which, by assumption, are all located in the SMM(brain), depend also on specific and appropriate ordinary matter brain activities.[4]

The 'single representation theory', which will be adopted here, assumes essentially that all mentality is represented by, and is as such transacted by, the SMM(brain). The ordinary matter brain acts only as a mediator (transducer) between the environment and regions of the ordinary matter body and the SMM(brain). The 'single representation theory' has an advantage over the 'double representation theory.' As far as SMM(brain) machinery is concerned, SMM(brain)s could generate and represent cognitive constructs, including memory traces, in the same way when attached to ordinary matter brains as when detached from them (and out of the ordinary matter body) during OBEs. The only difference between a SMM(brain) which is linked to an ordinary matter brain and a detached SMM(brain) (during an OBE) is the nature of the assumed input to the SMM(brain) in the two cases. A SMM(brain) which is linked to an ordinary matter brain is assumed to receive its input (normally) from the ordinary matter brain, via gravitons emitted by the ordinary matter brain. By contrast, a detached SMM(brain) is assumed to receive its input in the form of Shadow Matter signals derived from the environment. These Shadow Matter signals serve as encoders of psi-information. The possible machinery and modes of encoding of these Shadow Matter signals will be discussed later. (It will be argued also, in the sequel, that Shadow Matter signals could interact with the SMM(brain) when the latter becomes only locally, partly, detached from the ordinary matter brain, but remains predominantly attached to the latter.)

Also, the 'single representation theory' could, as stated already (p. 34), assume that SMM(brain)s, and not ordinary matter brains, form conscious experiences as byproducts (so-called epiphenomena) of some of their processes. Hence, according to this theory, in normal mental processes, as well as in OBEs, conscious experiences, when they occur, are always by-products of SMM(brain) activities. One avoids thereby a theory in which the ordinary matter brain and the SMM(brain) would each give complete and separate representation of the same cognitive constructs and their conscious concomitants (where present), as could be the case for the 'double representation theory.' Since, in fact people seem to have, normally, one 'I' (i.e. one personal identity experience), a theory that would imply simultaneous double consciousness may be discarded in favour of the 'single representation theory.'

It must be stressed that the latter theory remains fully within the discourse of mechanistic materialism, since, by assumption, SMM(brain)s are made up out of Shadow Matter which has material properties and which can be[5] related to string theory (see section 1.1). Thus, the view that ordinary matter brains are sufficient for explaining cognitive events completely, may be badly mistaken, and psi-phenomena suggest that this is the case. Yet, even within the present theory one can remain within the discourse of mechanistic materialism (as defined on p. 12). This view is not shared by many, perhaps most, parapsychologists. Many parapsychologists accept as a basic doctrine of their philosophy that parapsychology 'clashes with materialism.' This is simply not the case, if by 'materialism' one means 'mechanistic materialism.'

Not long ago I championed still the old view (Wassermann, 1978) which is, of course, upheld in every medical school and every psychology department. I still believe that the ordinary matter brain plays an important role as a mediator between, say, the ordinary matter body and the SMM(brain). Hence, brain-behaviour studies must remain of the utmost importance for understanding the ordinary matter brain machinery. (The latter machinery may be involved, normally, in triggering Shadow Matter representations of the world by means of the SMM(brain).)

The assumption that conscious experiences are by-products of the functions of SMM(brain)s and not of the functions of ordinary matter brains could explain also something else. When, as I assume, during an OBE some, or all, parts of a SMM(brain) leave the ordinary matter brain (and the ordinary matter body), then people may occasionally experience during the OBE what they report afterwards. In fact, some people are in an 'unconscious' state while their OBEs (which, of course, are experienced consciously) take place. The unconsciousness applies then only to normal sensory perception and normal thought processes. This is consistent with the view that SMM(brain)s and not ordinary matter brains are in normal perception, thoughts etc., the *seats of consciousness*. (I am leaving open here the likely possibility, that in some cases, during an OBE, only part of an SMM(brain), and not necessarily the whole of it, leaves the ordinary matter brain (and body).)

We could expect also that there occurs an exchange of information between the

SMM(brain) and the ordinary matter brain as long as that SMM(brain) remains normally bound to the ordinary matter brain. This exchange could be assumed to be reciprocal. Without this exchange it would be difficult to understand how information gained, say by forming memory traces, during an OBE could, after re-attachment of the SMM(brain) to the ordinary matter brain, enable people to talk about their OBEs. Since by assumption, Shadow Matter and ordinary matter only interact gravitationally, this suggests, as mentioned repeatedly, that gravitational quanta, so-called 'gravitons,' which are minute packets of gravitational energy (see Penrose, 1976a, 1976b), could serve as the means of communication between the Shadow Matter brain (SMM(brain)) and ordinary the matter brain. The communication would be in both directions, i.e. from ordinary matter brain to its bound SMM(brain) and vice-versa, and proceed with the speed of light.

The SMM(brain), when suitably activated, while partly, or wholly, attached to the ordinary matter brain, could emit gravitons which reach the ordinary matter brain, thereby triggering the ordinary matter brain into action. According to Eccles (1986) only minute amounts of energy might be required for triggering critically 'set' nerve cells of ordinary matter brains into action. I suggest, therefore, that graviton bombardment of critically set synapses of nerve cells (i.e. junctions of nerve cells) of ordinary matter brains could provide this triggering. This triggering could be analogous to the manner in which quanta of light (so-called photons) can trigger critically light-sensitive cells (e.g. cones and rods of retinas) of human and many animal eyes. Thus, just as retinal light-receiving cells have evolved as[6] photon-sensitive structures, so, I suggest, many central nervous synapses (i.e. nerve cell junctions), possibly all of them, may have evolved so that, apart from their other familiar functions, they are graviton-sensitive instruments.

1.5 Vision in 'Out of the Body Experiences' via 'Shadow Matter Eyes'.

Before I proceed with my further discussion of OBEs in relation to Shadow Matter and SMMs, I must first deal briefly with a supposed difficulty concerning OBEs, which was raised by Parker (1975, p. 104). He alleged that OBEs raise 'Some conflict with what is known about perception. So much of the external world we see is shaped and programmed by our perceptual systems that one could not conceive of an "eyeless vision",' which Parker believes occurs in OBEs. Yet, according to my theory, the postulated SMM(body) contains not only a SMM(brain). Among other things it could contain also SMM(body) regions that represent, in terms of Shadow Matter, the ordinary matter body's eyes and the optic nerve and so forth. Accordingly, the SMM(body) is assumed to be endowed with Shadow Matter eyes (see p. 15), linked to the SMM(brain) by Shadow Matter representations of the optic nerves.

Moreover, just as my theory assumes that any electron can become (reversibly) bound to a Shadow Matter selectron and each quark can become bound (reversibly) to a Shadow Matter squark (see p. 12), so I assume that any photon (i.e. a particle of light) can become bound (reversibly) to a Shadow Matter photon, called here a 'sphoton'. When a photon becomes absorbed by other ordinary matter, then it may separate from its companion sphoton and that sphoton may travel on. Likewise, if a sphoton, bound to a photon becomes absorbed by Shadow Matter, then the photon can move on.

Accordingly, if the percipient of an OBE is, say, in a room where various ordinary matter objects emit bound photon-sphoton pairs, then the Shadow Matter eyes could lead, via the Shadow Matter representation of the optic nerve, to activation of the SMM(brain) of the out of the body SMM(body) of the percipient. This could then represent the Shadow Matter analogue of normal vision. Hence, the OBE percipient could perceive his SMM(body), and the ordinary matter 'representations' of his surroundings, in a way analogous to normal vision. This could be so since the percipient's SMM(brain) could in an OBE, as in normal perception, provide normal representations for perception (including colour perception and 3-dimensional vision perception), memory traces and thoughts etc.

Also, if, as my theory assumes, during an OBE part or the whole of the percipient's SMM(body) occupies a location outside the percipient's ordinary matter body, then the view 'seen' by the Shadow Matter eyes would correspond exactly to the view seen by the percipient in normal perception if he occupied the out of the body location assumed to be occupied by the SMM(body) of the percipient during the OBE. Now, according to known case histories (see **Case 5**), the percipient's quasi-material body (see p. 29) can move about relative to the actual position of the percipient's ordinary matter body. Also, by assumption the SMM(body) represents the quasi-material body. Hence, the 'view' perceived by the SMM(body) of its surroundings could change continually during its OBE, in agreement with observations (see **Case 5**). I conclude that Parker's view that OBEs involve 'eyeless vision' does not apply to the present theory of psi-phenomena, at least not if one admits the hypothesis of the existence of Shadow Matter eyes (and other major assumptions of p. 29).

1.6 Further Properties of OBEs and possible Causes of altered states of Consciousness in Dreams, Mediumistic 'Trance States' and Hypnotic States.

Even at this stage of the book a great many, possibly important, theoretical insights have been extracted from information about OBEs. Accordingly, in this and the next few sections, I shall analyze OBEs much further, before proceeding to other topics.

In some near death OBEs, the percipients reported, according to Moody (1976; see also Sabom, 1982), that they were apparently unconscious, as far as normal perception was concerned. But during this period their OBEs had a striking sense of reality, which could not be mistaken for a dream. Some percipients reported that during the OBE they passed first through a narrow dark tunnel towards an opening, and, emerging there, they saw their own ordinary matter body, from the outside, in a location where their body was, in fact, sited (see Moody, 1976). Their body was, say, in a bed, or involved in a street accident, or located on an operating table (see Sabom, 1982 for OBEs during surgery), surrounded by people who tried to resuscitate them, or operate on them, and whose presence was later verified by the people concerned. In some, but not all, cases the OBE percipients reported also the presence of a very bright light as they emerged from the dark tunnel. This light persisted for a while. Yet, this light did not interfere with the OBE percipients' clear perception of normally coloured, three-dimensionally appearing, people and objects that surrounded their bodies. The light did not interfere either with the perception of people or objects which the percipients 'saw' during their OBEs in other nearby locations (say in a hospital) which they 'visited' during their OBE (see Sabom, 1982 and Moody, 1976). Before this, near the onset of the OBE, the percipients experienced, in some cases, a 'review' of earlier experiences in their life in great detail. The reviewed experiences occurred either in extremely rapid sequence, or were present in simultaneous display. Thus, many memory traces of some percipients were, apparently, activated spontaneously, either sequentially or simultaneously, during the OBE, and with a clarity that seemed to exceed ordinary memory recall (see Moody, 1976, pp. 49 ff).

According to my theory, the memory traces, which were activated during the OBE review, could be caused by hyper-activation of the SMM(brain) of the percipient. This hyper-activation could be due to the severing of many, previously tight, bonds between the SMM(brain) and the ordinary matter brain. This could result in increased autonomous activity of the SMM(brain). This could have several consequences. Among others it could lead to the increased clarity of memory recall and to the striking sense of reality, noted above. It could lead also to activation of many, normally dormant, memory trace systems of the SMM(brain). The SMM(brain) is, now, largely, or completely, detached from the ordinary matter brain, thereby enabling the SMM(brain) to give rise to the review. The state of the SMM(body) in which the latter is detached from the ordinary matter body could permit also direct access of incoming sphotons (Shadow Matter photons see pp. 36-7) to Shadow Matter eyes of the detached SMM(body) (see section 1.5 and Introduction p. 15). Sphoton activation of Shadow Matter eyes could lead, via activation of appropriate Shadow Matter representatives of the nervous system, to activation of the SMM(brain) (which forms part of the SMM(body)). If this interpretation is correct, then this could have important consequences.

According to my theory, telepathy and clairvoyance depend on Shadow Matter signals. Some of these signals could be sphotons (see p. 37), while others could be of a different nature to be discussed in chapter two. These latter Shadow Matter signals are assumed to be emitted directly by SMM(object)s (in particular SMM(brain)s) which are usually (but not in OBEs) bound to ordinary matter objects. When a Shadow Matter signal system, derived from a SMM(object) reaches a percipient, it could interact either (in the case of sphotons) with the percipient's Shadow Matter eyes (see p. 37) or with the percipient's SMM(brain). The Shadow Matter signal system could affect, therefore, directly or indirectly, the SMM(brain) of the percipient. I assume that the strength of binding of the percipient's SMM(brain) to the ordinary matter brain may decide whether the SMM(brain) reacts significantly to the clairvoyantly or telepathically derived incoming Shadow Matter signal system. In particular, I assume that tight binding between the ordinary matter brain and the SMM(brain) (which normally is assumed to be the case) inhibits telepathic and/or clairvoyant activation of the percipient's SMM(brain) by incoming Shadow Matter signals.

By contrast, states of weakened bonding (or partial or complete non-bonding, as in OBEs) between SMM(brain) and the ordinary matter brain facilitate autonomous action of the SMM(brain) as well as telepathy and/or clairvoyance by the percipient. (Dreams could result from intermediate weak bonding between SMM(brain) and ordinary matter brain, caused by that physiological state of the ordinary matter brain which produces also, indirectly, rapid eye movements (REMs) of the subject. In dreams the more weakly bound SMM(brain) could be more autonomous than in the waking state.)

In fact, if, while tightly bound to ordinary matter brains, SMM(brain)s could respond invariably to telepathic or clairvoyant input, then normal cognitive activities of SMM(brain)s could be interfered with constantly by telepathic (etc.) psi-information. It is plausible, therefore, to assume that tight coupling, as suggested, strongly inhibits telepathy and clairvoyance and other psi-experiences, such as OBEs. This suggests that ordinary matter brains may have evolved so as to ensure, among other things, normally tight coupling between the SMM(brain) and the ordinary matter brain. Alternatively, it could be simply a natural tendency of Shadow Matter to bind tightly, if possible, to appropriate ordinary matter of matching structure. Hence, whereas SMM(brain)s that are partly, or completely, detached from their matching ordinary matter brains could respond readily to telepathic or clairvoyant input, this might apply only rarely to SMM(brain)s that are tightly bound to ordinary matter brains.[7]

If these assumptions are valid, then they suggest that there may occur ordinary matter brain states that can lead to a loosening of tight binding of the SMM(brain). This could create states in which a SMM(brain), or a part of it, becomes weakly bound to the ordinary matter brain. In extreme cases there could occur partial, or complete, detachment of a SMM(brain) from the ordinary matter brain, leading to an OBE (when accompanied by appropriate detachment of part or most of the remainder of SMM(body) from the ordinary matter body). Some of the states of a weakly bound SMM(brain) could be associated with altered states of consciousness (e.g. dreaming, see above). Weakened bonding could apply particularly to those SMM(brain) states that may be involved in trance of trance mediums. Truly excellent trance mediumship, like that of Mrs Piper, is a rare, but not unique, phenomenon.[8] There was also, for instance, Mrs Gladys Osborne Leonard, who seemed to have extraordinary mediumistic powers according to the physicist Sir Oliver Lodge (1916) who attended some of her sessions.[9] Again, there was Mrs Eileen Garrett, whose mediumship seemed remarkable, particularly her communications concerning the details of the disaster of the *R101* airship (Price, 1931). Mrs Garrett, apparently, communicated information concerning intricate detail of the crash of the *R101* in 1930. The communication came ostensibly from the dead commander of the craft. The information was totally unknown to Mrs Garrett, but was corroborated as correct by living officials, and, hence, could, conceivably, have involved telepathic communication between Mrs Garrett's SMM(brain) while the latter was weakly bound to her ordinary matter brain, and the surviving SMM(brain) (which was not bound to an ordinary matter brain) of the commander of the airship. Alternatively, there could have been telepathic communication between Mrs Garrett's SMM(brain) and the SMM(brain)s of the living officials who had memory traces of the relevant information.

The fact that, apparently, able trance mediums, like Mrs Piper, Mrs Leonard and Mrs Garrett are very rare (and that many others, who claim to have mediumistic powers have them, often, at best, in very feeble forms) suggests that ordinary matter brain states (and/or SMM(brain) states) capable of generating good trance states are uncommon. This would not be surprising if, as my theory suggests, SMM(brain)s are normally (at least in the waking state) tightly bound to ordinary matter brains. Weak or weakened binding, say, due to complete breakage of many bonds between the ordinary matter brain and the attached SMM(brain) could, as in the case of trance mediums, be exceptional. The breakage of such bonds could be confined strictly to specific regions of the ordinary matter brain, while other ordinary matter brain regions remain firmly attached to the SMM(brain).

Ordinary telepathic or clairvoyant experiences of 'normal' people (other than trance mediums) could come about as follows. They could happen during very short occasional periods of weak binding of a part, or the whole, of their SMM(brain) to their ordinary matter brain, without these people developing proper trance states. I shall assume that in mediumistic trance, as well as in cases of spontaneous episodes of telepathy or clairvoyance, a substantial number of bonds between the ordinary matter brain and the SMM(brain) become broken. These bonds, however, can become re-established after cessation of the trance or spontaneous episode. The number of broken bonds would be larger for a mediumistic trance state than during occasional spontaneous psi-episodes or during dreams.

By contrast, during an OBE, all bonds between large parts, or the whole, of the ordinary matter brain and the SMM(brain) become (reversibly) completely severed. During restricted regional bond breakage those SMM(brain) regions that become locally detached from the ordinary matter brain (while remaining bound to parts of the SMM(brain)) could function completely autonomously, i.e. independently of the ordinary matter brain to which they are normally attached.

I have dealt now with one type of assumed SMM(brain)-bonding change. This kind of change could lead to altered states of consciousness. These altered states could occur during mediumistic trance or during spontaneous telepathy or clairvoyance and, perhaps, during dreams. I suggested how one could explain such altered states of consciousness in terms of my theory of psi-phenomena. I shall suggest now how essentially the same theory could give a partial explanation of the nature of hypnotic states. I assume that hypnotic states involve a state of affairs opposite to the 'broken bond' states between ordinary matter brain and SMM(brain) assumed in some psi-experiences.

By assumption in the hypnotic state some or many of the bonds between the ordinary matter brain and SMM(brain) become strengthened beyond their normal bonding strength. This could simply mean that many Shadow Matter constituents of the SMM(brain) move slightly closer to the ordinary matter brain constituents to which they normally bind tightly. The resulting strengthening of bonds between the ordinary matter brain and the SMM(brain) could lead to a state of the SMM(brain) where the latter becomes even more tightly bonded to the ordinary matter brain than normally. In consequence of this, during hypnosis, the SMM(brain) could be in a state where it becomes non-autonomous, because of its augmented bondage to the ordinary matter brain. In this state the SMM(brain) action, instead of being partly autonomous, could be entirely ruled by the ordinary matter brain. This could involve an abnormally high exchange of information-conveying gravitons between ordinary matter brain and the SMM(brain). This physical change of state (and loss of autonomy) of the SMM(brain) in hypnosis could involve an altered state of consciousness of the SMM(brain). The altered physical state of the SMM(brain) in hypnosis could lead to hyper-activation of the SMM(brain) by the ordinary matter brain, via the assumed increased graviton exchange. In other words, it leads to increased control of the SMM(brain) by the ordinary matter brain. This could lead, for example, to easier reactivation (recall) of existing memory traces stored by the SMM(brain).[10]

Let me illustrate how the preceding theory of the hypnotic state could be applied to some practical situations. Consider the following experiment, reported by Gindes (1953, p. 33):

A [hypnotized] soldier with only grade school education was able to memorize an entire page of Shakespeare's *Hamlet* after listening to the passage seven times. Upon awakening [from the hypnotic trance], he could not recall any of the lines, and even more startling was the fact that he had no remembrance of the hypnotic experience. A week later he was hypnotized again. In this state, he was able to repeat the entire passage without a single error. In another experiment, to test the validity of increased memory retention, five soldiers were hypnotized en masse and given a jumbled 'code' consisting of twenty-five words without phonetic consistency. They were allowed sixty seconds to commit the list to memory. In the waking state, each man was asked to repeat the code; this none of them could. One man hazily remembered having had some association with a code but could not remember more than that. The other four soldiers were allowed to study the code for another sixty seconds, but all denied previous acquaintance with it. During re-hypnotization, they were individually able to recall the exact content of the code message.

My theory of hypnosis offers the following explanations of these cases. Because of the physical change of state of the SMM(brain) during hypnosis, memory traces that are formed during that state could differ in structural ways from memory traces that are formed normally by the SMM(brain). The changed structure of the SMM(brain) during hypnosis could depend on abnormally high graviton input from the ordinary matter brain. According to my theory, such an augmented graviton input is available during hypnosis, but not in the normal state (see above). Hence, some memory traces formed during hypnosis could require for their reactivation (recall of memories) an augmented input of gravitons, which is present only during hypnosis. Hence, in the normal waking state the graviton input would be insufficient to reactivate the memory traces of the SMM(brain) formed during hypnosis, when the SMM(brain) was in its abnormal (hypnotic) state. This, then, could explain Gindes' experiments.

Nevertheless, suggestions made under hypnosis can have long-lasting post-hypnotic effects. This indicates that certain memory traces formed during hypnosis can have after-effects during normal (post-hypnotic) states of the SMM(brain). Possibly the structural change of the SMM(brain) during hypnosis does not affect all parts of the SMM(brain) equally. The parts of the SMM(brain) that store memory traces of hypnotic suggestions may not be the same as the parts that store memory traces of, say, poetry. (In fact, even those who, like myself (Wassermann, 1978), believed, or still believe, that ordinary matter brains are the exclusive means of storing memory traces, are agreed that different memories may be stored (in multiple copies) in different parts of the ordinary matter brain.)

The following case is instructive and illustrates the long-lasting post-hypnotic effect of a suggestion under hypnosis. The case illustrates also that the ordinary matter brain could, as I suggested, jointly with the SMM(brain), produce complex 'apparitions,' as a result of hypnotic suggestions, and that such apparitions can be even more intricate than that reported in **Case 1** (p. 24).

Case 3a

Gindes (1953, p. 39), who reported the case wrote:

It is truly remarkable how long... post-hypnotic control can last. In one of my experimental cases a college student was told under hypnosis that she would see her brother (who had been dead for two years!) six months later at 11am on a specified date. At precisely the appointed time, she was astounded to 'meet' her brother on a street in Los Angeles. These are her words: 'I was so happy to see him, but was astonished because I knew he was dead. However, in a way, his presence seemed perfectly natural at the time. He accompanied me to my apartment, and there we talked about different things, but in all that time, nothing was ever mentioned about his leaving. Soon he rose from his chair with the excuse that he had to keep an appointment, and left. I did not become fully aware of the impossibility of the situation until after he was gone, and then felt dazed. This feeling of bewilderment stayed with me until it was explained that my illusion was part of a hypnotic experiment.'

This case illustrates first of all the astonishing effects that hypnotic suggestions can produce, and the extraordinary power of phantasy. My theory attributes this power of phantasy to SMM(brain) action. **Case 3a** makes also more plausible the possibility that SMM(brain)s could produce complex apparitions in response to telepathic and/ or clairvoyant input signals. Various ostensible examples of this will be cited later in this book, and one example (**Case 1**, p. 24) was already given in detail. It might be tempting to suggest that OBEs are also entirely hallucinations produced by the highly creative SMM(brain) (or even by the ordinary matter brain.) This, however, becomes less plausible, when it is realized that OBEs involve often (but not always) some very accurate representations of verifiable, environmental circumstances. Hence, OBEs involve more than pure phantasy, generated either by the ordinary matter brain (according to conventional wisdom) or by SMM(brain)s while the latter are linked to ordinary matter brains. I shall argue below that OBEs involve not only separation of the bulk of the SMM(brain) from the ordinary matter brain (as already assumed on p. 13). OBEs may involve also strong hallucinations by the (out of the ordinary matter body) SMM(brain). Yet, I have argued repeatedly, that hallucinations alone could not suffice to explain OBEs, (see pp. 27-29) granted that, as in **Case 3a**, people can have immensely intricate hallucinations.

I conclude, at this stage, that there could exist a spectrum of states of SMM(brain) attachment to an ordinary matter brain. These states could range from abnormal extra-tight binding in hypnotic states, via normal tight binding in normal waking states, where this tight binding is assumed to inhibit responses to incoming psi-signals, to weak binding, and even complete detachment (in an OBE). Weak binding between SMM(brain) and the ordinary matter brain, which was assumed to occur in mediumistic trance, was attributed to the breakage of many bonds between SMM(brain) and ordinary matter brain. It must be stressed that the preceding interpretations do not exclude the possibility of a subject experiencing psi-phenomena during hypnosis. Let me remind the reader that, by assumption, hypnotic states are based on abnormally tight binding between SMM(brain) and the ordinary matter brain. Despite this, as in normal tight binding, some regions of the SMM(brain) could experience, in hypnosis, localized, and, hence, very restricted, breakage of bonds. This could result in localized weak binding of particular small sub-regions of the SMM(brain) to the ordinary matter brain. This should enable the weakly bound SMM(brain) 'patches' (if present) to receive incoming psi-signals during hypnosis.

Finally, I must note that although the preceding partial theory of hypnosis could explain some central aspects of the hypnotic state, it does not try to explain how particular ordinary matter brain changes (or possibly SMM(brain) changes) could lead to the strengthening of bonds between the two kinds of brains.

1.7 More about the Nature of 'Out of the Body Experiences' (OBEs).

I have assumed, above, that OBEs and mediumistic trance states and ordinary dreams all result from different degrees of breakage of bonds between the SMM(brain) and the ordinary matter brain. This is consistent with reports by Assailly (1963) that 'six out of his ten mediums reported OBEs' (cited from Parker, 1975, p. 110).

The assumed autonomy conferred on detached SMM(brain)s during OBEs[11] is consistent with reports that the cognitive faculties of near-death OBE percipients seem to be enhanced greatly. Moody (1976, p. 41) reports that 'Over and over, I have been told that once they become accustomed to their new situation, people undergoing this experience began to think more lucidly and rapidly than in physical existence.'

Also, in near-death OBEs visual perception seems to be much enhanced compared to normal perception (see Moody, 1976, p. 41), although objective tests are, obviously, not possible. By contrast 'hearing' in this state, according to Moody (1976, p. 42 and p. 47):

Can apparently be called so only by analogy, and most say that they do not really hear physical voices and sounds. Rather they seem to pick up the thoughts of persons around them, and. . . this same kind of direct transfer of thought can play on important role in the late stages of death experiences.

Moody continues:

As one lady put it. . . I could see people all around, and I could understand what they were saying. I didn't hear them audibly, like I'm hearing you. It was more like knowing what they were thinking, exactly what they were thinking, but only in mind, not in their actual vocabulary. I could catch it the second before they opened their mouths to speak.

This, if generally valid, provides direct evidence that near-death OBE percipients are capable of remarkable telepathic communication, according to my theory, between their partially, or wholly, detached SMM(brain) and other people's SMM(brain)s. This, of course, is consistent with the preceding assumption, that partial or complete abolition of binding of the SMM(brain) to the ordinary matter brain in an OBE should greatly enhance the power of the SMM(brain) to respond to telepathic and/or clairvoyant input (see chapter 2 for detailed mechanisms).

Readers might wonder at this stage why hearing with Shadow Matter ears should not be possible in OBEs, whereas vision with Shadow Matter eyes is possible (by assumption, see section 1.5). One reason for this is not difficult to fathom. Light is transmitted by light quanta, minute particles of light, called photons (see p. 36). These can have separable effects on a light sensitive nerve cell on the retina. By contrast, sound is propagated by the collective action of air molecules in the form of sound waves. Air molecules could be bound to corresponding Shadow Matter constituents. The latter would have to act collectively on the Shadow Matter ears of the SMM(body) during an OBE. This seems most improbable, unless the Shadow Matter ears are structures which function mechanically (in terms of Shadow Matter) like ordinary ears. Whether OBE percipients can 'pick up' music stored in the form of memory traces, or music that is listened to by other people, telepathically from other people's SMM(brain)s remains unknown.

I noted earlier that some people believe that OBEs are entirely hallucinated by the ordinary matter brain, which, in their opinion, transacts all mental processes. This, however, seems unlikely. The fact that percipients of OBEs give accurate accounts of what they saw and what happened during their OBEs, notably things they could not have seen from where their ordinary matter body was located, makes the *all* hallucination scenario, unlikely. So did my independent considerations (pp. 28, 33 and 47). Likewise Moody (1976) and Sabom (1982) rejected emphatically the 'hallucination only' hypothesis of OBEs on different grounds. The following case shows also, directly, why the 'hallucination only' assumption of the genesis of OBEs seems implausible.

Case 4

On page 32 of his book, Moody (1976) cites a lengthy report by a woman, who, during her OBE, felt herself drifting from her hospital bed past the light fixture at the ceiling. She stated:

> I watched them reviving me from up there! My body was lying down there stretched out on the bed, in plain view, and they all standing around it. I heard one nurse say, 'Oh my God! She's gone!' While another one leaned down to give me mouth-to-mouth resuscitation. I was looking at the back of her head while she did this. I'll never forget the way her hair looked; it was kind of short. Just then I saw them roll this machine in there, and they put the shocks on my chest. When they did, I saw my whole body jump right off the bed, and I heard every bone in my body crack and pop. It was the most awful thing!

It is somewhat difficult to see how the patient could have hallucinated her experience only by means of her ordinary matter brain. Particularly how she could have hallucinated the back of the head of the nurse who was giving her mouth-to-mouth resuscitation, while the nurse was doing so. Clearly, from her position on the bed the patient could not have seen, by normal vision, what she described she saw from her position at the ceiling during her OBE. It seems to me simpler to assume that the patient's SMM(body), including her SMM(brain), became detached from her ordinary matter body and moved to the ceiling. The SMM(body) could have remained connected to the ordinary matter body by the Shadow Matter cord discussed earlier. With the aid of Shadow Matter eyes the patient's SMM(body) could have seen her ordinary matter body from the outside, the short hair of the nurse, and what else she described, by the mechanisms given in section 1.5. In fact, **Case 5** suggests strongly, in harmony with my interpretation, that during an OBE some entity leaves the ordinary matter body. Moreover, the SMM(body) (including the SMM(brain)), when detached from the ordinary matter body during an OBE, could, analogously to a TV antenna dish, become re-orientated in space in optimal positions for receiving particular Shadow Matter signals, e.g. sphotons. This could facilitate the construction of an OBE which is spatially and temporally coherent and which corresponds to what took place actually.

Nevertheless, the following case shows that even if we assume a detachable SMM(body) (and SMM(brain)), then the detached SMM(brain) must be able to produce amazing phantasies, amounting to hallucinations, although these hallucinations do not create the OBE. This is not surprising, according to my theory. During hypnotically suggested post-hypnotic hallucinations the SMM(brain) could produce the most elaborate phantasies (see **Case 3a**, p. 42). Also, in any creative activity of gifted people a great deal of phantasy takes place. According to my theory, this is transacted by the SMM(brain) while attached to the ordinary matter brain. Possibly high creative activity could involve slight partial detachment of the SMM(brain) from the ordinary matter brain. This could allow a greater degree of autonomy to the SMM(brain).

The following case, like some of Moody's cases, suggests also that the SMM(body) is a highly elastic system that can be compressed or expanded readily in appropriate circumstances. This case shows also clearly that, apart from superimposed phantasy during an OBE, subjects can see surrounding objects and people from their apparent out of the body position with an amazing accuracy. In many cases what was seen during an OBE could be corroborated by independent witnesses. Also, the superimposed phantasy during an OBE is of a kind also present in dreams and ordinary hallucinations.

Case 5

The following case history comes from a paper by F. W. H. Myers in the *Proceedings of the Society for Psychical Research* (London) 1882, vol.8, pp. 180-94. The case shows, incidentally, like other cases cited by Tyrrell (1953, pp. 149-54) and others (see Gauld, (1977) for references), that such cases were known already last century. In fact, OBEs have been reported since antiquity; see Moody (1976, pp. 82ff). The case history concerns a physician, Dr A. S. Wiltse of Skiddy, Kansas. I shall rely on a mixture of the published report of the Society for Psychical Research and Tyrrell's

(1953) abridged version.

Dr Wiltse. . . lay ill with typhoid and subnormal temperature and pulse, felt himself to be dying, and said good-bye to his family and friends. He managed to straighten his legs and arrange his arms over his breast, and sank into utter unconsciousness. Dr S. H. Raynes, the only physician present, said that he passed four hours without perceptible heart-beat, although he could perceive an occasional gasp from Dr Wiltse. Raynes thrust a needle deep into the flesh of Wiltse at different points from the feet to the hips and got no response. According to Wiltse this state of near death lasted only half an hour during which he was absolutely unconscious. He then came again into a state of conscious existence and discovered that he was still in the body but the body and he had no longer any interest in common. He wrote 'I looked in astonishment and joy for the first time upon myself - the me, the real Ego, while the not me closed it upon all sides like a sepulchre of clay.'

One is here reminded of the passage through a 'dark tunnel' in several of Moody's (1976) OBE case histories; see p. 37.

Wiltse then described in minute detail how he 'watched the interesting process of separation of *soul* and body. By some power, apparently not my own, the Ego was rocked to and fro, literally, as a cradle is rocked, by which process its connection with the tissues of the body was broken up. After a little time the lateral motion ceased, and along the soles of the feet, beginning at the toes, passing rapidly to the heels, I felt and heard, as it seemed, the snapping of innumerable small cords. When this was accomplished I began slowly to retreat from the feet, towards the head, as a *rubber cord shortens.*' (Italics are mine.)

Before I continue the case history, a comment may be useful. Although Wiltse refers to a 'soul' or 'ego' which became separated, this is simply conventional mentalistic talk (see also my comments in the Preface). What matters is that Wiltse felt that apparently some entity became separated from his ordinary matter body (judging, at least, by his subjective report and in agreement with my assumptions). Accordingly, Wiltse's report is consistent with the view that during his near-death OBE his SMM(body), including his SMM(brain), became separated from his ordinary matter body. The reported 'snapping of innumerable small cords' is consistent with the breakage of my assumed gravitational bonds between SMM(body) and the tissues of the ordinary matter body (see p. 23), followed by an elastic contraction of the apparently compressible, SMM(body). This elasticity is hardly hallucinatory but a class characteristic of the phenomena. It occurs also in **Case 7**. But let me resume the case history of Wiltse:

I remember reaching the hips and saying to myself, 'now there is no life below the hips.' I can recall no memory of passing through the abdomen and chest, but recollect distinctly when my whole self was collected into the head, when I reflected thus: I am all in the head now, and I shall soon be free. I passed round the brain as if it were hollow, compressing it and its membranes, slightly, on all sides, towards the centre and peeping out between the structures of the skull, emerging like the flattened edge of a bag of membranes.

It can be seen that, apparently, consciousness resided in the system that was separating from the ordinary matter body (and, hence, ordinary matter brain). This is consistent with my assumption that the SMM(brain) and not the ordinary matter brain is the seat of consciousness. Moreover, if Wiltse's report were entirely based on a hallucination, then it would be surprising that he should hallucinate such a sequentially well-ordered withdrawal of the supposedly hallucinated entity from the ordinary matter body. This withdrawal could be explained readily by the actual withdrawal, by means of elastic contraction, of a material SMM(body). In agreement with the hypothesis of SMM(body) withdrawal one could assume that the SMM(body) has also Shadow Matter copies of the whole of the ordinary matter nervous system. This is consistent with my earlier assumptions that SMM(body) has Shadow Matter eyes (see section 1.5) linked to the SMM(brain) by means of Shadow Matter copies of the optic nerves. The Shadow Matter nervous system could then convey to the SMM(brain) the prevailing siting of the withdrawing SMM(body), in Wiltse's case, relative to the ordinary matter body, during withdrawal of the SMM(body) from the ordinary matter body. Thus, a consistent set of assumptions avoids the hypothesis that this, apparently very realistic sequence of events, as well as other phenomena discussed already, are nothing but, hallucinations. The assumed SMM(body) and its SMM(brain) and their behaviour could account for a host of minutiae (see p. 37) which the hallucination hypothesis explains only in the sense of attributing everything that happens to hallucinations. Explaining everything by hallucinations amounts in the end, hardly to an explanation at all.[12] I think, simply, that the assumed existence of a SMM(body) and its SMM(brain) and their behaviour fits the facts about OBEs well. Apart from this, the assumed SMM(body) could be consistent with the assumed existence of Shadow Matter, as suggested by string theory in physics (see section 1.1).

To return to the Wiltse case:

I recollect distinctly how I appeared to myself something like a jelly fish as regards colour and form. As I emerged, I saw two ladies sitting at my head...As I emerged from the head I floated up and down and laterally, like a soap-bubble attached to the bowl of a pipe until I at last broke loose from the body and fell lightly to the floor, where I slowly rose and expanded into the full stature of a man. I seemed to be translucent, of a bluish cast, and perfectly naked. With a painful sense of embarrassment I fled to the partially opened door to escape the eyes of the two ladies whom I was facing. . . but upon reaching the door I found myself clothed, and satisfied upon that point I turned and faced the company. As I turned, my left elbow came into contact with the arm of one of the two gentlemen who were standing in the door. To my surprise, his arm passed through mine without apparent resistance, the severed parts closing again without pain, as air reunites. I looked quickly up his face to see if he had noticed the contact, but he gave me no sign, only stood and gazed toward the couch I had just left. I directed my gaze in the direction of his, and saw my [ordinary matter] body. It was lying just as I had taken so much pain to place it, partially upon the right side, the feet close together and the hands clasped across the breast. I was surprised at the paleness of the face.

Some more comments at this stage. Much of what Wiltse saw during his OBE, as regards ordinary matter things, corresponded to the facts, as known to others. Yet, if one assumes a SMM(body) and its SMM(brain) to exist, then one must assume also that during the OBE the SMM(brain) produced, in this case, some major hallucinations while out of the body. Wiltse finding himself clothed suddenly was one such hallucination. Moreover, if Wiltse's SMM(body) contained, as suggested earlier, a Shadow Matter representation of the ordinary matter optic nerves, then one could expect also that the SMM(body) contains as well Shadow Matter representations of the rest of the nervous system. I assumed also that the Shadow Matter representation of the optic nerves mediates Shadow Matter signals between the Shadow Matter eyes and the SMM(brain). Similarly we could expect that the remainder of the Shadow Matter nervous system of the SMM(body) could also mediate Shadow Matter signals to, and from, the SMM(brain). So how could an ordinary matter body, like the arm of the man standing at the door (in the Wiltse OBE), pass through the Shadow Matter arm of Wiltse's SMM(body) without causing any pain? One possible answer is that neither the ordinary matter arm of the man standing at the door, nor the SMM(arm) bound to that arm, could interact with the Shadow Matter nervous system of Wiltse's Shadow Matter arm of his SMM(body) during the OBE.

So, back to Wiltse's report:

Wiltse tried to attract the attention of people in the room (e.g. by outright laughing), but without success. He notes that 'I then turned and passed out of the open door, inclined my head watching where I set my feet as I stepped down on to the porch. . . I crossed the porch, descended the steps, walked down the path into the street. There I stopped and looked about me. I never saw the street more distinctly than I saw it then. I took note of the redness of the soil and the washes the rain had made. . . Then I discovered that I had become larger than I was in earth life. . . My clothes, I noticed, had accommodated themselves to my increasing stature, and I fell to wondering where they came from and how they got on me so quickly and without my knowledge. I examined the fabric and judged it to be some kind of Scotch material, a good suit, I thought, but not handsome. . . The coat fits loosely too. . . Suddenly I discovered that I was looking at the straight seam down the back of my coat. How is this, I thought, how do I see my back? And I looked again to reassure myself, down the back of my coat, or down the back of my legs to the very heels. I put my hand to my face and felt for my eyes. They are where they should be, I thought. . . I turned about and looked at the open door, where I could see the head of my body in a line with me. I discovered then a *small cord,* like a spider's web running from my shoulders back to my [ordinary matter] body and attached to it at the base of the neck in front. (Italics are mine.). . . ' (Wiltse eventually returned to his ordinary matter body.)

What are we to make of some of this report? On p. 47 I cited from Wiltse's account. When he first appeared to leave his ordinary matter body he seemed to have a quasi-material body which looked something 'like a jelly fish,' was naked and 'translucent, of a bluish cast.' If this were hallucinated, like the suit of Scotch material a new question arises. If the ordinary matter brain could, as alleged by the 'hallucination only lobby,' hallucinate a suit as accurately as it did, and if it can allegedly produce details of human anatomy as accurately as in **Case 3a** (p. 42), then why should that brain hallucinate first the jelly like body of bluish cast in the Wiltse case? This suggests that the quasi-material body of bluish cast was not hallucinated, but genuine and a SMM(body). Again, when Wiltse first broke loose from the ordinary matter body, after falling on the floor he rose and expanded to the full stature of a man. On page 46 I gave Wiltse's description which suggests, in terms of my theory, that as the SMM(body) was trying to get out of Wiltse's ordinary matter body it began to 'retreat from the feet, towards the head, as a rubber cord shortens.' I deduce that the SMM(body) is elastically contractible. This conclusion is also reinforced by Wiltse's further statement (see p. 48) that after he crossed the porch he discovered that he had become larger than he was in ordinary life. This, again, is consistent with a further elastic expansion of the SMM(body). Yet, Wiltse also states that as he expanded his (hallucinated) clothes accommodated themselves to his increasing stature (p. 48). So we could have here a mixture of a non-hallucinated SMM(body) and an adjustable (variable) hallucination. Since, according to the present theory, during an OBE the SMM(brain) functions more autonomously when it is largely or nearly completely detached from the ordinary matter brain, this could facilitate the genesis of hallucinations by the SMM(brain). The capacity of hallucinated things or people to adjust themselves to given surroundings is not only demonstrated by the way Wiltse's hallucinated clothes accommodated themselves to his increasing size. It is shown in vastly more complex form by the way the apparition of the dead brother of the college student accommodated itself to the situation and the surroundings in **Case 3a** (p. 42).

It can be seen that in the Wiltse OBE, as in the cases reported by Moody and Sabom, there occurs a great deal of accurate information about the percipient's surroundings at the time of the OBE. Superposed on this were hallucinations (e.g. the suit of Scotch material above). All this is consistent with the view that the percipient's SMM(body), including his SMM(brain), became detached from his ordinary matter body and moved to various locations described. During his 'walkabout' the percipient's SMM(body) seemed to walk normally relative to the ordinary matter surroundings, directing his steps accurately relative to the ground. I mentioned repeatedly that in view of the amazing hallucinatory capacities (as in **Case 3a**, where the hallucinated dead brother of the subject walked also perfectly normally) the whole of the preceding OBE might conceivably be due to a hallucination. I gave already reasons why this is unlikely. The 'all hallucination' view of OBEs also raises other difficulties. A great many OBEs share very specific class characteristics (see p. 28), described by Moody (1976) (some are summarized on pp. 37-8). It seems unlikely that in a very large number of cases of OBEs, time and again, percipients should hallucinate precisely the 'correct' typical class characteristics of these phenomena. For instance, typically most or all percipients seem to look at their ordinary matter bodies from outside these bodies during OBEs.

The assumption that an OBE involves a SMM(body) and the latter's SMM(brain) which also participates in normal perception, thought and behaviour, could explain many more details of the OBE than the all-explanatory, vague, 'hallucinations only hypothesis.' This, of course, does not mean that the out of the body SMM(brain) cannot manufacture hallucinations. Indeed, the preceding **Case 5** and others (see **Case 7**) show clearly that many, perhaps all, OBEs involve some hallucinatory elements, which could be caused by the more unfettered activity of the SMM(brain) when freed from its bonding to the ordinary matter brain.

Case 5 shows also that Wiltse in his OBE was capable of normal colour perception (e.g. in the case of the red soil in the street outside the house). He was also capable of rational thought, decision making and memory formation, despite the fact that his ordinary matter brain was apparently functioning subnormally. This, again, suggests strongly that there could be a SMM(brain) which produces the most varied perceptions (e.g. of colour), produces thoughts and decisions (etc.) both in normal cognitive activities and in OBEs. This SMM(brain) can act independently of the ordinary matter brain during OBEs. During 'normal' states the SM(brain) is, as noted earlier, assumed to be tightly linked to the ordinary matter brain. According to my theory, when both 'brains' are linked they interact. How could the SMM(brain) act in colour perception during an OBE? I assumed previously that sphotons (i.e. Shadow Matter photons) are normally bound to photons (i.e. light quanta). When a sphoton-photon pair meets a Shadow Matter eye of the SMM(body) during an OBE then the sphoton could activate an appropriate component of that Shadow Matter eye. This, via the Shadow Matter optic nerve could act on the SMM(brain). The effect could be the same as if the photon of the sphoton-photon pair had acted on an ordinary matter eye, and in this way activated the ordinary matter brain, and, via the latter, activated the SMM(brain) while this is normally bound to the ordinary matter brain. According to the preceding theory it is, even during normal perception, the SMM(brain)'s activation which produced colour sensations, as mental by-products, and not the ordinary matter brain which does so.[13]

In particular, my interpretations are consistent with the assumption that during many OBEs a SMM(body) actually detaches from the ordinary matter body and visits localities described by the percipient after re-attachment. Also, the 'small cord' which in **Case 5** (and other cases) seems to link the percipient's quasi-material body to his ordinary matter body during the OBE is consistent with the view that the SMM(body) remains linked during an OBE by a Shadow Matter cord to the ordinary matter body. This Shadow Matter cord seems to be elastically extensible and contractible. It is here believed to be ultimately instrumental, by elastic contraction, in reuniting the SMM(body) with the ordinary matter body.

The alternative assumption that there exists no SMM(body) and no SMM(brain), but only the ordinary matter body and its brain, would require a vast amount of continuing coherent clairvoyant information during the OBE, and one would then have to explain also how that clairvoyance works without Shadow Matter. For instance, if Wiltse had not left his body, how could he know that his face looked as pale as it appeared (see p. 47), or that other objects and people were just where they were. Or, why should the small cord be hallucinated in **Case 5** and in various other cases reported in the literature of the subject? Within the present Shadow Matter theory the 'small cord' plays a significant role. Also, the assumed Shadow Matter eyes and the SMM(brain) etc., could explain the continuity and intrinsic coherence of the experienced, continually changing, aspects of the OBE. In fact, many (but not all) of the experienced aspects of the perceived environment (during an OBE) would be exactly those experienced by an actual human ordinary matter body and its sense organs, etc. if present in the varying localities 'visited' during the OBE. These veridical OBE experiences differ drastically in kind from the non-veridical experiences of hallucinations, e.g. the post-hypnotic hallucination reported by Gindes (**Case 3a**).

1.8 Interpretations of two somewhat more unusual 'Out of the Body Experiences'.

The preceding theory suggests that OBEs could be explained by assuming that part or most of the SMM(body), together with most, or all of its component SMM(brain), becomes detached from the ordinary matter body during the OBE. It is assumed that the detached part (or sometimes the whole) of the SMM(brain), while detached, could produce a good deal of hallucinatory material. The following case history seems to be unique in some respects. It stretches the preceding interpretation to its limits. The case, reported by Tyrrell (1948, p. 195), was described to Tyrrell by a lady well known to him. He had complete confidence in the accuracy of her account. It occurred in August 1921.

Case 6

The lady states:

I was lying in bed cogitating about doing something extremely agreeable but entirely selfish. I was suddenly aware of being in two places at once. One 'me' was still lying in bed looking as I normally do. The other 'me' was standing at the foot of the bed, very still, very straight, dressed in white with a Madonna-like veil over the head. I was aware of the extreme whiteness of the clothes. We then had a spirited discussion. The white 'me' said: 'You know you will not do this.' The 'me' in the bed flung itself about in exasperation at the impassive authority of the white 'me' and said: 'I shall do what I like you pious white prig.' There was no sense of a third 'me' linking the two. Each 'me' could see the other, with its expected exterior surroundings all the time. The white 'me' felt sympathy, but contempt for the other 'me'. I may say that the white 'me' won. I have no memory of the process of coalescing, merely at a given moment both 'me's were observing the exterior world from the same place.

Tyrrell noted, correctly, that this case involves a 'personality division,' a phenomenon familiar from cases of multiple personalities (see McDougall, 1944, chapter 31 for several case histories).[14] So, before I interpret the OBE, just quoted, let me cite extracts from McDougall's (1944, pp. 497ff) account of the multiple personalities of the Beauchamp case, reported by Morton Prince (1906), a medical man, in a nearly 600 page study. The subject was a nervous impressionable child. In addition to her normal personality B she developed three distinct personalities B1, B3 and B4. Of these B3 was also known as 'Sally' because of the name which that personality adopted. McDougall (1944, p. 498) noted that:

For nearly one year. . . B1 and B4, led the life of alternating personalities with reciprocal amnesia; and careful study of them during this time showed that they were complementary characters, each having command of the memories of the first period [before either of them emerged]. . . B1 was a humble weakly invalid, very suggestible, shy, retiring, considerate of others and fond of children and old people. B4 was very self-assertive, given to quick violent anger, intolerant and quarrelsome, vain, sociable, irreligious, disliking children and old people. There were corresponding differences in tastes. Both were very emotional, but whereas B1 was wholly swayed by her emotions, B4 fought them down. B1 was easily tired and relatively inactive, though studious. B4 was energetic and fond of bodily activity; she disliked most of the things that B1 liked. Yet another personality B3 turned up, that of 'Sally,' who 'was an impish, childish personality and showed remarkable consistency, without any indications of increasing maturity throughout the several (some six) years of her active career.' B3 manifested herself when B1 was in hypnosis (Prince referred to B1 in hypnosis as B2), 'speaking of B1 as 'she' and of herself as 'I,' and claiming to be a personality as entirely distinct from B1 as was possible under the circumstances, the circumstances namely that they inhabited and made use of the same bodily organism. The hypnotic state personality of B1 was very different from 'Sally'.

McDougall (1944, p. 500) cites Prince's summary of the relations between B1 and 'Sally' thus: 'Sally is a distinct personality, in the sense of having a character, trains of thought, memories, perceptions, acquisitions, and mental acquirements, different from those of B1. Secondly, she is an alternating personality in that during the times when the primary self [B1] has vanished Sally is for the time being the whole conscious personality, having taken the place of the other... At such times B1 does not become a subconsciousness to Sally but as a personality is wiped out [or rather is latent]. Thirdly, Sally does not simply alternate with B1. There are times when Sally manifests herself as an extra-consciousness, concomitant with the primary personality B1, while B3 became repressed.'

Prince described also that with the emergence of personality B4 the latter strove to suppress personality B3 (i.e. 'Sally'). The evidence marshalled by Prince suggests that the original personality B of Miss Beauchamp became 'split' into the alternating personalities B1 and B4. Finally Dr. Prince succeeded in recombining B1 and B4 into the single personality B.

The Beauchamp case belongs to abnormal psychology and not to parapsychology, although it has, I believe, an important bearing on the interpretation of psychic phenomena adopted here. Cases of multiple personalities suggest that whereas the ordinary matter brain of a person is a single coherent system, the SMM(brain) could consist of a, normally integrated, set of sub-SMM(brain)s, which can become dissociated. Accordingly, a SMM(brain) could become decomposed into sub-SMM(brain)1, sub-SMM(brain)2, sub-SMM(brain)3... etc. The sub-SMM(brain)s could normally cohere within the whole SMM(brain) like the pieces of an assembled jigsaw puzzle cohere within the whole puzzle. Each of the sub-SMM(brain)s, when active, is, by assumption, capable of having some of its activities accompanied by conscious experiences. Some of these experiences may be feelings of an 'I,' i.e. of a particular personality being consciously active. As long as the sub-SMM(brain)s of the total SMM(brain) cohere, there would exist only a single 'integrated' experienced 'I'.[15]

With this interpretation of multiple personalities at hand, I shall now return to the OBE of **Case 6** and try to explain it. In this OBE the lady-on-the-bed had the feeling of an 'I' (or 'me' as she put it) i.e. a 'personal identity feeling' which I shall call PIF1 for short. In addition, the lady-in-the-white-dress experienced a feeling of a second 'I' (or 'me') which I shall call PIF2. I shall assume that prior to the OBE, PIF1 was represented by, say, sub-SMM(brain)1 (and that this remained the case during the OBE). Also before the OBE, PIF2 was represented by sub-SMM(brain)2 and remained represented by it during the OBE. Both sub-SMM(brain)s were, prior to the OBE, bonded to the lady-on-the-bed's ordinary matter brain. I suggest that in this case (and, likewise with appropraite changes of detail, in the case mentioned on p. 32) the OBE involved a splitting of the lady-on-the-bed's SMM(brain) as follows. Sub-SMM(brain)1 remained attached to the lady-on-the-bed's ordinary matter brain, while sub-SMM(brain)2 together with the lady-on-the-bed's SMM(body) (of which it formed a part) became detached from her ordinary matter body and moved outside that body. The OBE was then experienced by sub-SMM(brain)2, which, via the Shadow Matter eyes of the SMM(body), 'saw' the actual ordinary matter body of the lady-on-the-bed. The Shadow Matter mechanism of seeing could be of the kind suggested already and discussed in connection with other OBEs (see section 1.5 and also p. 50). In addition sub-SMM(brain)1 could have remained bound to the ordinary matter brain of the lady-on-the-bed during the OBE (see above). The bonding of sub-SMM(brain)1 to the ordinary matter brain could have become subnormally weak (as in mediumistic trance, see p. 39), owing to the absence of sub-SMM(brain)2, which normally, by interacting with sub-SMM(brain)1, could help to increase bonding of the latter to the ordinary matter brain. When weakly bound to the ordinary matter brain, sub-SMM(brain)1 would produce the hallucinated figure of the lady-in-white, much as in **Case 5** Wiltse's SMM(brain), while completely detached from the ordinary matter brain during his OBE, hallucinated the Scottish material suit of the SMM(body) (by assumption). In addition there could have occurred telepathy between sub-SMM(brain)1 and sub-SMM(brain)2. This could have resulted in the experienced dialogue between the hallucinated lady-in-white and the out of the body experienced lady-on-the-bed. In fact, the hallucinatory 'body' of the lady-in-white could have been related to the SMM(body) of the lady-on-the-bed's, while that SMM(body) was out of the lady-on-the-bed's ordinary matter body.

It is, of course, tempting to ask: If sub-SMM(brain)1, while bound to the lady-on-the-bed's ordinary matter brain can produce such a powerful hallucination as the lady-in-white and her speech, then why should a SMM(brain) bound to an ordinary matter brain not be able to produce a complete OBE? Yet, the phantasmal figure, which was hallucinated, did not correspond to any environmental givens in that situation, and that figure (the lady-in-white) apparently saw the lady-on-the-bed from the outside. Also in the Wiltse OBE (**Case 5**), apart from the hallucinated suit of Scottish material, the percipient perceived all the correctly positioned people and scenery, and much of this was simply not hallucinated in the sense in which the lady-in-white was hallucinated. I conclude, therefore, again, that although hallucinations occur in OBEs, it seems unlikely that OBEs are entirely hallucinatorily constructed. The OBE-generating machinery, proposed here, could explain many minutiae of the class characteristics of OBEs. The following OBE case shows also, clearly, that an OBE may involve a host of veridical minutiae of the percipient's environment. It may involve also powerful clairvoyance (or telepathy as in Case 6).

Case 7

This case comes from Myer's (1892, p. 194) paper. The percipient was the Rev. L. J. Bertrand, a Huguenot minister. He gave Dr Hodgson, a prominent and very meticulous case investigator of the Society for Psychical Research, an oral account of this experience, and then sent a written report to the distinguished psychologist William James. I shall quote only some of the lengthy report:

Bertrand reached a group of pupils with whom he climbed, at the inn of the Engstelalp (Switzerland) - foot of the Titlis, 45 minutes from the Col of the Yoch. He writes:

While taking my meal on the grass, surrounded by companions, I said to the guide: 'Why do you not climb the beautiful Titlis straight up from this side instead of going round to meet the long zigzagging Trübsee Alp way?'... In spite of warnings from another guide (Karl Infanger) who assured the party that, because of prevailing snow and of an enormous bump of the glacier it would be perfect madness to risk the climb, Bertrand insisted on the climb to be undertaken. After Bertrand had walked from six in the morning until five in the afternoon he felt the strain in his legs. So he decided to stay behind and ask his companions to make the further ascent to the peak with the guide and asked the guide 'to go up by the left and come down by the right,' since on the left there was a dangerous cut in the mountain. Bertrand then sat down, as he writes 'my legs hanging on a dangerous slope or precipice, my back leaning on a rock as big as an armchair. I chose that brink because there was no snow, and because I could face better the magnificent panorama of the Alps Bernoises.' While lighting a cigar he 'suddenly felt as thunderstruck by apoplexy' and though the match burned his fingers he could not throw it down. 'My head was perfectly clear and healthy, but my body was as powerless and motionless as a rock... This,' I thought, 'is the sleep of the snows! If I move I shall roll down in the abyss, if I do not move I shall be a dead man in 25 or 30 minutes. My feet and hands were first frozen, and little by little death reached my knees and elbows. The sensation was not painful and my mind felt quite easy. But when death had been all over my body my head became unbearably cold, and it seemed to me that concave pincers squeezed my heart, so as to extract my life. I never felt such an acute pain, but it lasted only a second or minute, and my life went out. Well,' thought I, 'at last I am what they call a dead man, and here I am, a ball of air in the air, a captive balloon still attached to earth by a kind of elastic string up and always up.[16] How strange! I see better than ever, and I am dead. . . where is my last body?' Looking down I was astonished to recognize my own envelope. 'Strange' I said to myself, 'there is the corpse in which I lived and which I called me... What a horrid thing is that body! - deadly pale, with a yellowish-blue colour, holding a - cigar in its mouth and a match in its burned fingers! . . . Ah! if I only had a hand and scissors to cut the thread which ties me still to it! When my companions return they will look at that and exclaim "The Professor is dead." Poor young friends! They do not know that I never was alive as I am and the proof is that I see the guide up rather by the right, when he promised me to go by the left. . . Now the guide thinks that I do not see him because he hides himself behind the young men while drinking at my bottle of Madeira. . . Ah, there he is stealing a leg of my chicken. Go on old fellow, eat the whole chicken if you choose, for I hope that my miserable corpse will never eat or drink again.' I felt neither surprise nor vexation; I simply stated facts with indifference. 'Hallo!' said I, 'there is my wife going to Lucerne, and she told me that she would not leave before tomorrow, or after tomorrow. They are five before the hotel at Lugern. Well, wife, I am a dead man. Good-bye.'

Here we can see clearly that Bertrand in his out of the body state could exert full cognitive faculties. He could evaluate situations, presumably with the help of his detached SMM(brain), since his ordinary matter brain seemed unconscious (and, in any case is not the producer of conscious experiences according to the present theory). In fact, he could not, with the help of his ordinary matter eyes and brain have seen, from where he was located, many of the details and things that he noticed during his OBE. For instance he could not have seen the guide while hiding behind the young men drinking from his bottle of Madeira. In fact, in agreement with the preceding interpretations of other OBEs, the present case history suggests the following. The SMM(brain), while partly or fully detached from the ordinary matter brain, is freely accessible to Shadow Matter signalling systems. These systems can act on the SMM(brain) either directly (if the signalling systems are of the 'right kind,' see below) or via Shadow Matter eyes. This, for instance, could initiate clairvoyant representations of the nearby environment.

For perception of nearby ordinary matter objects in an out of the body state, the SMM(brain) could, possibly, rely on sphotons (Shadow Matter photons), emitted jointly with partner photons (see section 1.5) by nearby ordinary matter objects. These sphotons could be received by the Shadow Matter eyes of the detached SMM(body). From there, as explained already in section 1.5, the Shadow Matter nervous system could then activate the SMM(brain). Yet, sphotons are unlikely to convey accurate information to Shadow Matter eyes about objects many miles, in some cases thousands of miles, from the Shadow Matter eyes. The reasons are the same as those that exclude photon-mediated vision of distant objects, except of very intense light sources, such as distant stars. However this analogy between sphotons and photons might be wrong. If, however, each sphoton is emitted only jointly with a linked photon, then the analogy could hold. Just as light from relatively weak sources gets totally absorbed before travelling very far from the sources, so the same could apply to sphoton-based material energy. Moreover, without magnifying instruments (e.g. telescopes or binoculars) human ordinary matter eyes cannot discriminate much detail of the retinally imaged light of very distant objects. There are no obvious candidates for Shadow Matter equivalents of telescopes.

For these reasons I prefer to postulate a different mechanism for both telepathic and clairvoyant communication, at least in addition to possible short-range sphoton-mediated clairvoyance discussed above. This additional mechanism assumes that any ordinary matter object whatsoever, inanimate or animate, forms first of all, a primary SMM(object), much as a human ordinary matter body is assumed to form a SMM(body). The SMM(object) is assumed to remain attached normally to the object by gravitational bonds between SMM(object) and object. These bonds are between Shadow Matter and ordinary matter. The SMM(object), in turn, is assumed to be able to act as a template for the frequent formation of Shadow Matter models that are complementary to a part, or the whole, of the SMM(object). These complementary SMMs will be called SMM(complem)s.

After its formation a SMM(complem) is assumed to become ejected from the SMM(object) that served as its template. This would make room for the formation of another SMM(complem) by the same template. In this way, an ordinary matter object could, via its firmly attached SMM(object), act as a source of frequent formation and emission of SMM(complem)s. SMM(complem)s are assumed to be as elastic and expansible or contractible as SMM(objects)s (see footnote, p. 68 and pp. 47-8 and p.60).[17]

I shall denote by SMM(complem, body) any SMM that is complementary to SMM(body). A corresponding notation will be used for other SMM(complem)s. In particular SMM(complem, brain) is assumed to be capable of autonomous intellectual activities (as suggested in section 1.4, see particularly p. 34), especially capable of forming new memory traces. Also, when a SMM(complem, body), after separation from its template SMM(body), moves about freely, then the Shadow Matter eyes of SMM(complem, body) could be stimulated by sphotons (Shadow Matter photons). These sphotons could be emitted by SMM(object)s bound to objects which that SMM(complem, body) encounters while moving about freely. The sphotons, by stimulating the Shadow Matter eyes of SMM(complem, body) could then lead to the formation of memory traces by the SMM(complem, brain) of that SMM(complem, body). In this way the SMM(complem, brain) could form memory traces of objects encountered by SMM(complem, body) while moving about. This, as I shall argue in section 2.4, forms the partial basis of my theory of clairvoyance.

Ordinary vision is due, in parts, to photons that are emitted by a host of objects surrounding an ordinary matter body of a person (or animal). These photons lead to the indirect activation of that person's ordinary matter brain. This, by assumption, leads to the activation of the SMM(brain) of that person. Clairvoyant perception, as just explained above, leads to the activation of SMM(complem, brain)s of SMM(complem, body)s and this could lead to the formation of memory traces about the surrounding objects or events, etc.

A complete theory of clairvoyance requires further assumptions. These will be put forward in section 2.2.1 and 2.4. But let me stress the following point. In ordinary vision a percipient receives photons from distant or near sources. Yet, the percipient does not necessarily move towards the source of photon emission. By contrast, in the present theory of clairvoyance SMM(complem, body)s, emitted by a percipient (or replicated in a way that will be explained later) can move to all sorts of near or far objects, and receive sphotons emitted by these objects when they are sufficiently close to them. What happens subsequently will be explained in chapter 2.

At this stage, I must return to, and round off, the Bertrand case (**Case 7**). We left Bertrand on p. 55. His further thoughts can be found in Myer's paper and are very interesting as regards the complexity of thoughts during the out of the body state. They suggest that, if my theory is valid (see pp. 34-5), then the detached SMM(brain) is capable of producing, on its own, complex, and emotionally finely shaded thoughts, and is capable of producing also new memory traces, while detached. (It is also interesting that, according to Bertrand, his quasi-material body had no hands and was ball-shaped (see pp. 54-5 and also p. 29 for and explanation of a similar case.)

Late in his account Bertrand wrote:

Suddenly a shock stopped my ascension, and I felt that somebody was pulling the balloon down. My grief was measureless. The fact was that whilst my young friends threw snowballs at each other our guide had discovered and administered to my body the well-known remedy, (rubbing with snow), but as I was cold and stiff as ice, he dared not roll me for fear of breaking my hands still near the cigar. I could neither see nor hear any more, but could measure my way down, and when I reached my body again I had lost hope - the balloon seemed much too big for the mouth. Suddenly I uttered the awful roar of a wild beast - the corpse swallowed the balloon, and Bertrand was Bertrand again, though for a time worse than before.

Bertrand then described to the guide that he ascended by the right to the Titlis, and allowed the young men to put aside the rope. He then accused the guide of drinking from his bottle of Madeira and that he stole one of his chicken legs. [A roast chicken in those days was an expensive delicacy.] As Bertrand writes 'This was too much for the good man [the guide]. He got up, emptied his knapsack whilst muttering a kind of confession, and then fled away.' Also Bertrand's wife had stopped at the Lugern Hotel with five people and had gone to Lucerne, as experienced by Bertrand during his OBE (apparently clairvoyantly, say by the mechanisms of section 2.4).

1.9 'Testability' of the theory and the variety of its special assumptions.

Although the present theory invokes several special assumptions (technically known as *ad-hoc* hypotheses) for explaining each particular case history, this is really typical of most, if not all, scientific theories. I explained this elsewhere (Wassermann, 1989). In a typical scientific hypothetico-deductive theory, such as Newtonian Mechanics, one encounters hundreds, if not thousands of systems to which the theory has been applied. In applying the general theory to any particular system one requires particular assumptions special for that system only. In this case one says that the system is being 'modelled' by the general theory. For instance, in Newtonian Mechanics one may assume as an elementary system, that one has a 'smooth inclined plane' with a 'point-particle' sliding down a line of greatest slope of the plane. Or a point particle may slide down a smooth, thin, circular wire located in a vertical plane. In this way one gets one system after another for which system-specific assumptions are being made, in addition to the general 'laws' of mechanics (i.e. Newton's three 'laws' of motion). There is therefore nothing unusual when I introduce system-specific assumptions into my theory of psychic phenomena, in addition to general mechanisms which are used to explain many cases, and are applied to many different systems.

One way of testing my integrated theory of psi-phenomena is analogous to the way in which one has 'tested' many theories in physics. This is by using these theories to *explain* many already known phenomena, and by examining how these theories fit the phenomena. For instance, in 1905 Einstein explained the already known photo-electric effect, but did not predict it. His explanation also 'tested' Planck's earlier assumption of the existence and properties of light quanta (i.e. photons). Thus, Einstein explained the photo-electric effect by making assumptions about the properties of photons. Likewise, the theory of superconductivity of Bardeen and his associates (e.g. Cooper)[18] explained the already known phenomena of superconductivity, but did not predict them. Again, Felix Bloch established a well known quantum theory of electronic conduction in metals, which explained many already known facts and laws (e.g. Ohm's Law) without predicting them. Einstein, Bardeen and Bloch (etc.) obtained Nobel Prizes for their respective explanatory feats listed, showing that explanations are highly valued in science, at least as much as valid predictions. Successful explanations serve as strong tests of a theory even if (like predictions) they cannot prove a theory to be true.

My theory, as it stands, explains many already known phenomena of parapsychology, and is, thus, severely tested by these explanations. Another theory-linked test of my theory would be to demonstrate that some of its assumptions about Shadow Matter follow ultimately from a future *Theory of Everything* (see, p. 21) or are, at least, consistent with it. If so, as noted before (Wassermann, 1988, and p.11) psi-phenomena could provide the strongest indirect evidence for the assumed existence and properties of Shadow Matter (as postulated in physics). In this way, by harmonizing with a *Theory of Everything* of theoretical physics, psi-phenomena could help towards an important enlargement of the realm of mechanistic materialism, by giving Shadow Matter a firm footing.

I think that the dated view that 'testing' of scientific theories amounts always exclusively to verification (or else falsification) of 'predictions' of theories, can be discarded (see also Wassermann, 1989 and Wassermann, 1974, section 3.12). A scientific theory can be tested often by demonstrating the validity of its explanations as well as its predictions (if these exist). A good many years ago I quoted Professor Hans Jürgen Eysenck (Wassermann, 1974, p. 123). Eysenck (1953, p. 235) argued that it is 'not *ex post facto* explanations which constitute science, but predictions which can be verified.' Yet, following Eysenck, the retrospective explanations given by Einstein's theory of the photo-electric effect, or given by Bardeen *et al*'s theory of superconductivity, or given by Bloch's theory of electronic conduction in metals (see above), and many other explanations provided by many other scientific theories on a retrospective basis, form essential aspects of science. Eysenck, who is a prominent psychologist, started his career by studying physics. Surely he must know that explanations play a central part in physics and many other sciences?[19]

1.10 Could Psi-Phenomena be revealed by 'repeatable' Experiments?

Highly structured OBEs, 'apparitions' (see **Case 1**) and many other kinds of psi-phenomena occur spontaneously and sporadically, and, apparently, cannot be produced to order. I suggested (pp. 38-9) that psi-phenomena can only be registered by SMM(brain)s while the latter are either weakly bound to, or partially or completely detached from ordinary matter brains. If so, then it is futile to expect that psi-phenomena can be produced on demand by any normal person. This person's SMM(brain) could be bound so tightly to his or her ordinary matter brain, as to prevent perception of received psi-signals (see p. 38). To demand that psi-phenomena should always be repeatable to order, in specially designed, controlled, experiments seems to me totally unjustified. This demand is made often, allegedly so as to 'conform to the canons of science.' These canons supposedly demand such repeatability (see Beloff, 1970, p. 130). In fact, good science demands nothing of the sort (see Wassermann, 1955). *The kind of repeatability that scientists ask for is a repeatability of class characteristics of phenomena.* Scientists do not necessarily request repeatability of minutiae of experimental outcomes, unless these minutiae happen to be class characteristics of the particular phenomena investigated. I have noted already, extensively, in the case of OBEs that many of their general class characteristics turn up repeatedly, but spontaneously and sporadically (i.e. unpredictably) for different people (see pp. 28-9 and section 1.6). Thus, in one case, and in some others, a particular class characteristic may turn up, but in other cases this may not be so. Similar remarks apply to numerous other classes of psi-phenomena yet to be discussed. For instance, Tyrrell (1953) classified apparitions, and found specific class characteristics for different types of apparitions. For instance, some apparitions, as in **Case 1**, are collective apparitions which are simultaneously experienced by several people. Most other apparitions are only experienced by a single percipient.

One should regard case reports of spontaneously occurring, and well substantiated, psi-phenomena of a specific class as resembling, say, human medical case histories of a particular type of illness, say case histories of paranoid schizophrenia. Medical case histories, like case histories of psi-phenomena, vary in precise detail from case to case. They share also, for the same illness, striking common class characteristics, which make diagnosis possible for each particular class of illness. For ethical reasons one cannot reproduce human disorders, say the results of virus infections, to order, and for many human illnesses there exists no corresponding animal model. Again, one cannot produce cases of paranoid schizophrenia experimentally, for ethical reasons. (An amphetamine-induced psychosis may resemble paranoid schizophrenia. But it just is not paranoid schizophrenia.) This does not contradict that case histories of paranoid schizophrenia, with many or all of the class characteristics of the illness, turn up repeatedly, spontaneously and sporadically in human populations. What is repeatable about case histories of a particular type of illness are, therefore, its class characteristics (even if they are not all present in each particular instance of the illness). Similar remarks apply to particular kinds of psi-phenomena. To expect that psi-phenomena should turn up repeatably to order in experimentally imposed controlled conditions seems therefore, unreasonable (contrary to Randi, 1991 p. 88). It is as unreasonable as to expect new cases of tuberculosis to turn up to order, or that lightning, earthquakes or volcanic eruptions should turn up, when looked for, in the wild at any time.

The fact, therefore, that systematically controlled experiments in parapsychology have yielded only mostly negative results, or mostly (but not entirely) unimpressive data, must not distract from the fact that psi-phenomena occur spontaneously. I am disregarding here some very 'impressive experimental results' that turned out to be due to data manipulation by the experimenter (see Markwick 1978). Of course, one cannot exclude the possibility that people may discover drugs which detach partially the assumed SMM(brain) from the ordinary matter brain to an extent that psi-phenomena can occur, more or less, at one's command.

Typical, i.e. genuine, psi-phenomena cannot, currently, be produced to order on an impressive scale (with rare exceptions). This shows, incidentally, that those who claim to be able to produce such phenomena impressively on television or stage, at any time, rely invariably on trick procedures. Of course, the fact that professional conjurers can simulate certain aspects of psi-phenomena, when desired, does not mean that genuine psi-phenomena do not occur spontaneously. Some gifted conjurers, alas, have claimed that because their artifacts can simulate psi-phenomena, that all evidence for psi-phenomena in nature must be based on artifact or fraud. It is like saying that because bacteriological warfare could produce a plague epidemic, the medieval plagues must have been produced by bacteriological warfare, and could not have arisen spontaneously by the spread of infection.

I conclude also that repeatability to order of statistically highly significant results, in stereotyped card-guessing, etc., experiments in parapsychology, without the aid of possible psi-facilitating drugs, seems (with rare exceptions) to be at variance with what we know about psi-phenomena. These phenomena occur in their most impressive and highly structured forms, spontaneously and sporadically. It is, therefore, not surprising that some positive results of most systematic experiments in parapsychology were explained away, ingeniously, as possible artifacts of sensory cues, or possible fraud, etc. (see Hansel, 1980; Marks, 1986; G. R. Price, 1955). These explanations, however, were more often than not, of the variety of 'what could happen if pigs could fly.' A case of actual data manipulation turned out to be more prosaic (see Markwick, 1978). In any case, experimental parapsychology with its attempts to produce phenomena repeatably to order, has suffered from a dearth of sustained positive results, except for minor, and questionable, effects (see Beloff, 1977) (with perhaps a few rare but important exceptions).

There could be physical reasons why psi-phenomena cannot (at present) as a rule be produced to order (see p. 59). The 'experimenter effect' is often invoked to excuse negative results, as due to the experimenter's powers to inhibit (unconsciously) the psi-faculties of the experimental subject. This seems to me to be a highly questionable excuse.

It is for the preceding reasons that I prefer to rely in this book on case histories of spontaneous cases of psi-phenomena, of the kind that have repeatable class characteristics. The argument of certain philosophers, who are ignorant of science, that psi-phenomena do not belong to science because they are not 'repeatable' is absurd. It rests, as noted, on a profound misunderstanding of what, precisely, is repeatable in science, namely that it is class characteristics of phenomena that are repeatable. These philosophers would not dismiss the study of genetically predisposed disorders as unscientific simply because these disorders turn up sporadically in populations. Some of these 'critical' philosophers believe that repeatability in physics must be of the simple kind that occurs approximately in very simple experiments (e.g. to test Ohm's law or Boyle's law) that are performed in schools. They forget that there are many experiments in elementary particle physics where thousands if not millions of (automated) observations are required until a looked for effect turns up.[20]

1.11 Hyperscepticism and Prejudice vis-a-vis Psi-phenomena.

Scientific prejudice, notably against parapsychology, is essentially no different from any other kind of prejudice (say against Jews or coloured people). Its explanation must be looked for in the general psychology of prejudice (see Allport, 1954). It is estimated that around 30% of any population, are prone to go with established fashion. Such individuals are usually inflexible, with obsessively conservative attitudes as regards new discoveries, new ideas, the unfamiliar and things they cannot explain (Allport, 1954). Often no amount of evidence or arguments will sway these people. For example, the great German physicist von Helmholtz, famed already in the nineteenth century, proclaimed authoritatively that 'Neither the testimony of all the Fellows of the Royal Society nor even the evidence of my own senses could lead me to believe in the transmission of thoughts from one person to another, independently of the recognized channels of sensation' (see Flew, 1953, p. 120 or Polanyi, 1951). It did not occur to von Helmholtz, prejudiced as he was, that there could be future discoveries in physics and other sciences, revealing types of matter and their interactions, which he never dreamed of. Many such new discoveries did, indeed, turn up.

In the early days of investigation of hypnotism (then termed 'mesmerism' after Mesmer, who rediscovered hypnosis), virulent prejudice against hypnotic phenomena was as common as prejudice then, and later, against psi-phenomena. To illustrate the attitudes that prevailed then vis-a-vis hypnotic phenomena, it is useful to sample old issues of the medical journal *Lancet*. Thus, in *Lancet* of July 1843, vol.2, p. 534 it is stated that:

> The mesmerists have recently been attacked in the pulpit at Liverpool by that noted declaimer the Rev. Hugh Mcneil, who directly charges them with being the agents of Beelzebub, if they produce any effects on patients, though he doubts that mesmerism is anything else than fraud, not possessing the importance which necessarily belongs to dealings with Satan. If it *be* a science he demands that its laws be demonstrated until which he cautions all Christians to avoid lectures by Mesmerists.

(For other comparable passages see *Lancet* 6 December, 1845, vol. 2, p. 629; 20th June 1846, vol. 1, p. 689; 10th August 1850, vol. 2, p. 181; 14th June 1851, vol. 1, p. 653; 23rd May, 1857, vol. 1, p. 553; 4th February 1843, vol. 1, p. 686.)

The end of most, or all, opposition to the use of hypnosis for medical treatment (e.g. in dentistry) did not come until near the middle of the twentieth century (at least in the United Kingdom). Tyrrell (1948, p. 229) noted the drastic change of view and stated that:

In the early days, hypnotism (then called mesmerism) called for bitter opposition and was regarded as being on a par with paranormal phenomena. The committee appointed by the Society for Psychical Research to report on hypnotism quoted *Lancet* as saying: 'We regard the abettors of mesmerism as quacks and impostors; they ought to be hooted out of professional society.' The medical profession in those days refused to admit the genuineness of hypnosis. When the most painful surgical operations were successfully performed in the hypnotic state, they said that the patients were bribed to sham insensibility; and that it was because they were hardened impostors that they let their legs be cut off and large tumours cut out without showing any signs of discomfort. At length this belief, in all but the most bigoted partisans, gave way before the triumphant success of Mr Esdaile's surgical operations under hypnosis in the Calcutta Hospital. . . (Tyrrell cited a large part of this passage from the *Proceedings of the Society for Psychical Research* (London) vol. 2, pp. 154ff)

Indeed, similar gross prejudices have been common about parapsychology. As the eminent British journalist Bernard Levin (1980, p. 124) put it, many people 'display the kind of panic reaction that I predicted for Brian Inglis's *Natural and Supernatural*, and that, when the book was published, fulfilled my prediction ninety times over and then nine more. (The book is a history of the paranormal from the earliest times to the first World War. I predicted a terrified rejection of the evidence by those unable to face the facts that there are things in the universe they are unable to understand.)'

This, however, does not mean that we must, or should, accept uncritically all the numerous, often gullible claims made by blind believers in everything that is claimed by certain people concerning the paranormal (see West's 1954, p. 19 many relevant critical remarks). Great care and expertise is required to separate genuine phenomena from fraud and humbug in parapsychology. To my mind, for instance, some of James Randi's sceptical demonstrations on TV are hopelessly inadequate.

Some, obviously absurd, conjurers have argued that psi-phenomena could be genuine only if they occurred when demanded under conditions specified by the conjurers. This, however, cannot be demanded for spontaneous and sporadic phenomena, such as psi-phenomena. It would be like demanding that an eclipse of the sun must occur at a time prescribed by a conjurer. Scientists simply cannot dictate to spontaneously occurring natural phenomena (such as lightning, earthquakes and psi-phenomena) when, and in which fashion, they should occur. For instance, animal speciation events in the wild seem to occur spontaneously and unpredictably. Yet, nobody would deny that there exist species.

One reason why scientists accept readily certain rare events (such as lightning, earthquakes etc.) is because they can be explained in terms of accepted scientific theories. Until recently (Wassermann, 1988) nobody had put forward a workable scientific theory explaining how mechanisms, linked to current fundamental physics, could explain spontaneous and sporadic (or experimental) psi-phenomena.

The purpose of this book is, of course, to make further progress along the line of establishing a theory which goes much beyond my earlier, provisional theory

(Wassermann, 1988) and which may link psi-phenomena via Shadow Matter theory to a physical *Theory of Everything* (see p. 21 section 1.1). Anti-psi scientists, and others, notably some philosophers, believed that physicists had already discovered all conceivable major components of matter and all forces of nature. They believed, therefore, that there are no more possible means available for explaining psi-phenomena via physics. Indeed, at more than one conference on parapsychology I have heard proclamations that physicists had discovered fully all the basic ingredients of nature as regards particles and forces. This book suggests that such an outlook was, probably, as badly premature as the views of many nineteenth century sceptics regarding the genuineness of hypnotic phenomena. I cited already Dr Beloff's statement (Introduction, p. 16) 'that we must abandon hope of a physical explanation of psi-phenomena,' and that we must rely, instead, on some theory based on an immaterial soul. Beloff, a former president of the Society for Psychical Research (London) has not presented such a theory. I believe that if, as this book suggests, it is feasible to explain psi-phenomena in terms of an appropriate Shadow Matter physics, then few scientists, if any, will share Beloff's view that psi-phenomena require an immaterial soul. By assumption, Shadow Matter has distinct physical properties (e.g. it can become bound to appropriate ordinary matter) and this does not apply to the immaterial soul (but see Preface concerning a material 'soul').

The philosopher Flew (1953, pp. 120ff) has cited a substantial list of alleged reasons why psi-phenomena, if genuine, would 'clash' with then known scientific theories. Like all prejudiced people, Flew, could not anticipate subsequent striking advances in theoretical physics. He could not anticipate string theory (see pp. 21-2), and the possible existence of Shadow Matter linked to string theory, let alone the possibility that Shadow Matter might be able to explain all major psi-phenomena. Alas, there exists a long line of philosophers, dating back to the days of Galileo (and before) who made pronouncements about what is and what is not going to be possible in science. Many of these pronouncements have become the stuff of jokes.

Many of those laymen, who are largely or totally unfamiliar with the contents of the many volumes of the *Proceedings* and the *Journal of the Society for Psychical Research* (London) and a host of other scholarly works on psi-phenomena, as well as many academics, tend to dismiss psi-phenomena as 'the result of mere chance coincidence' without any meaning. Yet, which natural phenomena are meaningful lies in the eye of the beholder, and the existence of class characteristics is crucial. If that beholder is a theoretical physicist, like myself, who feels tempted to make sense of putative psi-phenomena by interpreting them in terms of the notions of theoretical physics, then this seems to me in order. The fact that we do not call rare observations, which result from high energy particle collisions, meaningless coincidences, is due to a body of theory which allows us to interpret the observations in a meaningful way. Thus, whether we attribute meaning to rare natural phenomena or regard these as 'mere chance coincidences' depends on whether the phenomena make sense within an acceptable theory, or whether we have no such theory. In fact, the primary purpose of this book is to make sense of many psi-phenomena, which sceptics have hitherto regarded as mere chance coincidences or as hallucinations or as contradicting physics, by interpreting these phenomena within a framework that may be related to string theory in physics.

Another type of hyperscepticism comes from conjurers (see my remarks on p. 60 and p. 62). Some of these claim to be able to simulate any putative psi-phenomena by appropriate trick procedures or to show that in their presence the phenomena do not happen (see Randi's (1982) views, which I do not share). Even if these claims were entirely valid, they do not entail that psi-phenomena are necessarily produced by conjuring tricks or do not exist (Randi should have tried to explain how OBEs are produced by conjuring tricks! But then he would, presumably, fall back on the unacceptable 'hallucination hypothesis.') Now, computer simulations of a physical process is not the same as the, say, naturally occurring physical process which the computer program simulates. Likewise a conjurer's trick-based simulation of a very limited range of psi-phenomena need not be the same as the psi-phenomena that are simulated, and it need not imply that these psi-phenomena are produced by tricks.

To sum up, my past conversations with academic hypersceptics have convinced me of the following. No amount of evidence for psi-phenomena, even if witnessed by highly competent observers, would satisfy that fraternity, irrespective of how many investigators or reporters of repute supply the evidence. They will always argue that one or two notorious case histories of spontaneous cases turned out to be based on mistaken memories, etc., and that, therefore, all case histories must be suspect and rejected. Likewise, no type of theory, however closely linked to physics or other sciences, could shift the hardened prejudice of such people. They are as prejudiced in their views of psi-phenomena as the venomously anti-hypnotism lobby was last century concerning excellent demonstrations of hypnotic phenomena. Perhaps some of these hypersceptics would make excellent material for serious students of the psychology of prejudice, even if their views on psi-phenomena are gratuitous.

Perhaps also something should be said about an organized group of hypersceptics called 'the Committee for Scientific Investigation of Claims of the Paranormal' (CSICOP) founded in 1976. They discussed, for instance, psychic phenomena in a British television programme *Equinox* (*Superpowers*? Channel 4, 6th October, 1991). This group included conjurer James Randi, a co-founder of CSICOP (see Randi, 1991, p. 13), who, judging by his writings seems to me to believe that there are no paranormal powers. This is so because psi-phenomena, being spontaneous, cannot be demonstrated in the unrealistic conditions (i.e. non-spontaneous conditions) in which Randi tries to demonstrate these phenomena 'to order' on television (see Randi (1991). *The Radio Times* (North East Edition of 5-11 October 1991) writes on p. 53, concerning the *Equinox* programme:

> An international group of sceptics wishes that people would not believe in ghosts... or telepathy. Their organization's mission is to stem the paranormal tide and defend science...

Surely, science can look after itself without the help of this group of sceptics, some of whose intellectual limitations are most unlikely to assist science? Worse for CSICOP, if psychic phenomena, such as telepathy and apparitions (or 'ghosts') should form part of the subject matter of science, because they can be linked with physics in terms of a coherent theory (and contrary to Randi, whose scientific credentials I do not know), then most, if not all, of these hypersceptics are barking up the wrong tree. Their claims sound like those of people who might claim that they must 'stem the tide' of quantum mechanics and relativity theory, because they might be used in understanding how nuclear weapons work, and, thus, harm humanity. Above all, I believe that hypersceptics fail to recognize repeatable class characteristics of psychic phenomena, where these demonstrably exist (e.g. Tyrrell's (1953) classification of apparitions).

1.12 Repeatable Class Characteristics of a Particular OBE.

Although spontaneous OBEs are not rare among psi-phenomena, I wish to show that some of the class characteristics of **Case 5** (p. 45) are also shared by at least one other OBE. Otherwise it could be argued that Case 5 is unique and not a member of a sub-class of OBEs. The case history comes from Muldoon and Carrington (1969) and is cited by Moss (1976, pp. 295-6).

Case 7a

Muldoon wrote that in one of his OBEs he [felt he] was floating in the very air, rigidly horizontal, a few feet above the bed. . . Then he found [his quasi-material body] standing upright in the room, but six feet above the bed. . . He was able to get a standing position on the floor however. Then he turned toward the bed he had been in and notes that:

There were two of me! I was beginning to feel myself insane. There was another 'me' lying quietly on the bed! *My two identical bodies were joined by means of an elastic-like cable,* one end of which was fastened to the medulla oblongata. . . I attempted to open the door, but found myself passing through it. Going from one room to another I tried fervently to arouse the sleeping occupants of the house. I clutched at them, called to them, tried to shake them, but my hands passed through them as though they were but vapours. I started to cry. . . As I recall it, I prowled about for perhaps fifteen minutes when (the cable) pulled. I began to zigzag again under this force. . . It was the reverse procedure of that which I experienced rising from the bed. Slowly the phantom [i.e. the quasi-material body] lowered, vibrating, then it dropped suddenly, coinciding with the physical body.

In **Case 7a**, as in **Case 5** (the Wiltse case) an elastic cable (or 'cord' in the Wiltse case) turns up, and links the quasi-material body and the ordinary matter body. Again Muldoon's hands passed through the people he tried to contact as if they but vapours, while Wiltse experienced that the arm of his 'quasi-material body' 'was passed through. . . without apparent resistance' by the arm of one of the people present, and that 'the severed parts' of the arm of the 'quasi-material body' closed again without pain 'as air reunites' (see p. 47). Also **Case 7a** again reports that the connecting cord or 'cable' is elastic. This is consistent with my repeated arguments that the 'quasi-material body' consists of highly elastic Shadow Matter (see p. 49, 50 and fn 16 p. 68).

1.13 Physicists predict a central Role for Heavy Shadow Matter in Astrophysics and Cosmology.

Some people might argue that the only evidence for the existence of Shadow Matter is, according to this book, the existence of psi-phenomena. This, however, is not the case. Kolb *et al* did not introduce Shadow Matter in connection with parapsychology. They argued on the basis of super-string theory that Shadow Matter could exist on purely mathematical grounds (see section 1.1), and then argued at great length that it could solve some major unsolved problems of Astrophysics and Cosmology.

Kolb *et al* envisage that apart from ordinary matter there exists a relatively *heavy* Shadow Matter, which may 'mirror,' particle type for particle type, the elementary particles of ordinary matter, notably in weight. By contrast, I have, in addition, postulated an immensely *light* Shadow Matter which can get bound to all ordinary matter objects which exist in the ordinary matter universe (see p. 13)

The heavy Shadow Matter could according to Kolb *et al* explain numerous problems in Astrophysics, etc. If there were roughly equal components of heavy Shadow Matter and ordinary matter in the disk of our Galaxy then this 'would explain one of the several dark matter problems that the gravitational mass of the disk (as inferred from dynamics) seems to be about twice that of the material we can see or detect (for example, stars, white dwarfs, gas, dust) [Bachall. J. N. (1984) *Astrophysical Journal, 286* 169-81]. One might also expect some binary systems comprised of an ordinary star and a shadow star. Such a system would manifest itself as an isolated star with periodic proper motion. In fact there are nearby stars. . .which are suspected of having invisible companions [Mihalas, D and Binney, J. (1981) *Galactic Astronomy, 45,* Freeman, San Francisco]. Of course, there are less exotic explanations for these invisible companions; they could be neutron stars, black holes, black dwarfs or planetary size objects.'

Kolb *et al* also argue 'that: it seems unlikely that the solar system would contain a significant amount of [heavy] Shadow Matter.' This, however, does not mean that there could not be a significant amount of light Shadow Matter, postulated here, in the solar system. As regards the Earth, Kolb *et al* argue that the heavy Shadow Matter (they consider) would settle at the centre of the Earth (which, again, does not apply to the light Shadow Matter). Whereas for each atom of the Earth there is assumed to exist a corresponding light Shadow Matter atom (called a satom), this does not apply to the heavy Shadow Matter constituents. Again, Kolb *et al* maintain that 'if there is only a small amount of [heavy] Shadow Matter in the Sun. . . then the Shadow Matter will sit at the centre of the Sun. . .' This again, need not apply to the light kind of Shadow Matter.

So the argument goes on and Kolb *et al* consider various other effects which heavy Shadow Matter could have on our universe. While, according to my theory, light Shadow Matter has effects on various systems in the universe which may differ drastically from the effects of heavy Shadow Matter, the possibility that heavy Shadow Matter exists inspired me to envisage that light Shadow Matter could also exist and have the effects explained in this book. Hence, whenever in this book, I refer to Shadow Matter I mean as before, unless otherwise stated, light Shadow Matter.

Notes to Chapter One

1. Every scientific theory, when applied, requires additional subsidiary assumptions. For instance, in very elementary mechanics one may treat a planet as if it were a 'point particle' (see p. 21). One may consider a 'perfectly smooth sphere,' moving down an inclined plane, and so forth. Such 'idealizations' are typical of elementary and advanced scientific theories, and they form subsidiary assumptions, which are specific for each case, and which do not form parts of the general aspects of a theory.

2. Celia Green (1976 p. 113) cites a similar case of a hospital patient who one morning felt herself floating upwards and found herself looking down on the rest of the patients and could see herself 'propped up against pillows, very white and ill.'

3. This assumption differs drastically from the 'mirror universe model' of Kolb *et al.* (1985), who assumed that Shadow Matter particles have the same masses as their corresponding ordinary matter particles.

4. That ordinary matter brain machinery plays an important part in memory trace formation in normal states (when the ordinary matter brain and the SMM(brain) are assumed to be tightly linked) is suggested by evidence that place-learning in rats is accompanied by long-term potentiation of hippocampal synapses (Morris *et al* 1986; Goddard, 1986). These findings, of course, do not show that ordinary matter brain components are the sites of learning trace formation. In fact, memories formed during OBEs are consistent with the view that SMM(brain)s are the sites of engram formation in OBEs and in normal engram formation (engram = memory trace) when ordinary matter brain and SMM(brain) are tightly coupled (by assumption). (See also Stevens (1989) and Bliss (1990).)

5. The present theory could apply also, subject to appropriate changes, to lower species that have suitable ordinary matter brains, and which could have SMM(brains)s bound to these ordinary matter brains.

6. In fact it has been estimated that a retinal light-receiving cell can be activated by a single photon. (For the underlying biochemistry see Wald 1981).)

7. If with 'normal' people (as distinct from trance mediums and other psychics) states of weakened bonding between SMM(brain)s and ordinary matter brain are intermittent and rare events, then this could explain some common findings. Many parapsychologists have conducted experiments in which a percipient had to guess a target known, or knowable, to the experimenter, but unknown to the percipient. Most of these experiments gave only results that one would expect by chance, and indicated that psychic faculties were not at work. But occasionally a particular subject would show a sustained run of guesses above chance expectation level. But after this run terminated, which was, usually, soon enough, no further abnormal scoring by the percipient occurred. This is consistent with the view, upheld here, that every now and so often the bonding between SMM(brain) and the ordinary matter brain of normal people can become weakened, so as to allow the SMM(brain) to respond to psi-input. If during such a period of weakened bonding a subject is tested for his or her capacity in guessing targets by paranormal means, the subject may score significantly above chance. But when this period is over, and tight bonding is resumed, the subject cannot score any more abnormally highly. If this interpretation is valid, then it is futile to search for experiments which will yield indefinite repeatability of a particular psi-guessing behaviour.

8. See Tyrrell (1948); Douglas (1976, Chapter 7); Gauld (1977); West (1954) and Piper (1929) for various aspects of the Piper trance mediumship.

9. For a brief account see Douglas (1976), Chapter 9, who described the Leonard mediumship far beyond the boundaries of the Lodge encounters with Mrs Leonard; see also her autobiography Leonard (1931).

10. Just an aside for physicists. That subtle changes of states between interacting systems can have drastic consequences is known from the theory of superconductivity. In superconduc-

tivity the conduction electrons of the super-conducting metal interact with each other and with the ionic lattice of the conductor. Apart from this the electrons can interact also pair-wise with each other via the ionic lattice. The latter interaction becomes dominant in superconductivity, when a change of state occurs, with lowered temperature.

11. I shall suggest, below, that during an OBE not necessarily the whole of a SMM(brain) may become detached completely from the ordinary matter body. Part of the SMM(brain) may stay behind and remain bonded to the ordinary matter brain. Hence, when I refer to the detached SMM(brain) during an OBE, I mean that part of the SMM(brain) that becomes completely detached from the ordinary matter brain and ordinary matter body and leaves the latter.

12. See also section 1.2 pp.27-9, section 1.4, p. 33 and see p. 37 and p. 42 for arguments against the hallucination view of OBEs and for facts that ought to be explained.

13. I must add a technical note. Neuro-psychologists might be tempted to argue that electro-stimulation of the human ordinary matter brain can elicit colour sensations. They could conclude, wrongly, that therefore it must be the ordinary matter brain which produces these sensations (see Penfield and Rasmussen, 1950). This, however, does not follow. Electro-stimulation of the ordinary matter brain could lead to activation of the SMM(brain). It could then be the SMM(brain) activity which, as a mental by-product (a so-called epiphenomenon) leads to the experienced colour sensation(s).

14. Some of McDougall's long accounts of typical case histories of multiple personalities are based partly on accounts given in William James' *Principles of Psychology,* in a paper by Rev. W.S. Plumer, DD. in *Harpers Magazine,* May 1860, in a book by Sidis and Goodhart (1905) and in Prince's (1906) book.

15. It might be objected to the theory of multiple sub-SMM(brain) composition of SMM(brain)s that ordinary matter brains could cater partially for activation of all memories (etc.) in terms of nerve cell junctions (so-called synapses). In the present theory, however, the SMM(brain) is assumed to be the seat of memory traces and can also form new memory traces when out of the body. If there are several separate personalities associated with the same person then each of these personalities could have its separate memories encoded by a personality-specific sub-SMM(brain).

16. The elastic string in this case is strikingly similar to the 'cord' which seemed to connect the percipient's experienced 'quasi-material body' (see p. 29 for this notion) and his ordinary matter body in the Wiltse case (see p. 50 and my comments p. 46 concerning the elasticity of SMM(body) and the possible elasticity of Shadow Matter in general). Also the 'elasticity' of the string in the present case reminds one of the apparent 'elasticity' of the presumed SMM(body) in the Wiltse case (see pp. 47-8). Also the upward floating of the apparent quasi-material body in the present case is similar to the drifting to the ceiling in Case 4, p. 44.

17. For readers versed in molecular biology it will be obvious that my theory is closely modelled on genetic transcription and translation in the following sense. Just as the DNA double helix (a so-called duplex) can serve as a template for forming complementary messenger RNAs, so a SMM(object) could serve as a template for forming SMM(complem)s. Just as messenger RNA, after completion, can become removed from its DNA template, so a SMM(complem) could be ejected from its SMM(object) template.

18. The late Professor Herbert Fröhlich contributed a key idea to the current theory of superconductivity (see his obituary in *The Times* (London) 30.1.1991 p. 16).

19. Falsifications à la Popper are something negative, valid explanations are constructive.

20. For instance, in the proton-antiproton collider experiments at CERN (Geneva) designed to discover the theoretically predicted W and Z intermediate vector bosons of electro-weak interaction theory, a vast number of observations were required to discover evidence for the expected very rarely occurring appearance of W and Z bosons (see Kalmus, 1983; Duff, 1986). Rubbia and Van der Meer received Nobel Prizes not only for their outstanding

techniques, but also for their patience in persisting with the experiments to discover these bosons.

2
Machinery
For
Psychic
Phenomena

2.1 The Relation between an Ordinary Matter Brain and its attached Shadow Matter Brain

2.1.1 How Coherent Shadow Matter Models (SMM(body)s) of Ordinary Matter Bodies could recognize and match accurately Ordinary Matter Bodies

By assumption, towards the end of an OBE the detached out of the body SMM(body) and its SMM(brain) re-enter the ordinary matter body, and reunite with it. To do so, the SMM(body)'s sub-components must be able to recognize, and match precisely, the ordinary matter body's sub-components down to the level of individual cells. This could enable many unique sub-components of the SMM(body) to attach to uniquely specified cells of the ordinary matter body. How is this possible? I shall give one possible answer to this question in terms of a grossly simplified version of a biological theory which I developed quite independently of parapsychology (Wassermann, 1972, 1978; Clowes and Wassermann, 1984).

This theory tries to explain how, during development from a fertilized egg into a mature body, myriads of cells recognize each other often with great accuracy. In many cases a cell seeks out, with the utmost accuracy, its chemically optimally matching partners. According to this theory most bodily cells could become so highly specified that they become individually recognizable, like the unique fingerprint of a particular individual. Such a theory was first suggested in the 1930s by Nobel Laureate Roger Sperry of the California Institute of Technology. Following in Sperry's footsteps, I drove his original theory a good deal further. I suggested how specific 'genetic programmes,' which a human (or other) fertilized egg inherits, could ensure, during development, by systematic gene actions, that a host of bodily cells become uniquely specified (Wassermann, 1972, 1973, 1978; Clowes and Wassermann, 1984). For instance, it seems probable that, just as the fingerprints of no two people in the world are the same, so the specification of no two typical brain cells, of the same individual, is the same (Wassermann, 1986). This is so because genes, in ways that are now understood at least in principle (see Clowes and Wassermann, 1984), could, during development of the organism, ensure systematically this astonishing degree of specificity.

Indeed, much evidence suggests that ordinary matter brains, and perhaps many other parts of ordinary matter bodies, are specified genetically, so accurately that for many organs and structures (e.g. bones) most organ-specific cells and structure-specific cells are chemically uniquely labelled. This has immediate consequences for my theory of psi-phenomena. By assumption, ordinary matter human bodies serve as templates for the formation of matching SMM(body)s (see pp. 12-3). Normally the SMM(body) remains attached to the ordinary matter body. Each cell, as it forms during development of the organism, could form a matching SMM(cell). Hence, as the ordinary matter body develops, so the SMM(body) and its SMM(cell)s develop correspondingly. But the ordinary matter body and its cells are genetically determined (subject to environmental influences). Accordingly the SMM(body), through its accurate formation by the ordinary matter body is also accurately determined, genetically, by the same genes that determine the ordinary matter body. In particular, the SMM(brain) is genetically accurately determined via the ordinary matter brain to which it is normally attached, both during development and in the natural state. I have simplified somewhat. In reality, during development of an organism, there occurs an interplay between environment and gene actions. The environment contributes also to the formation of an organism. The outcome, however, may still be strongly gene-dominated, as is shown by the amazing structural similarities of identical twins (technically called monozygotic twins). Yet, since the SMM(body) is, by assumption, formed by the ordinary matter body, in all its developing stages, the SMM(body) is produced automatically by the combined indirect effect of genes and environment. In addition, environmental Shadow Matter could also have an effect on the developing SMM(body).

My theory of development explains how it could be that no two organ-specific cells of the same organ (e.g. a brain),[1] or structure-specific cells of the same structure (say a thigh bone) are alike. This tremendous diversity of cells of, say, the brain, could be achieved by combinatorial labeling (see Clowes and Wassermann, 1984 for the machinery). This could ensure that each organ-specific cell of an organ is labelled uniquely, by precisely programmed genetic machinery. Likewise, each structure-specific cell of a bone becomes labelled uniquely in this theory. Indeed, if we think of the many fine protuberances and curvatures which a particular bone has to form, during its development, in all the correct places, just where and when required, then this suggests that every bone cell 'knows,' by means of 'chemical cues,' just the right place where it has to be located at the right time. The problem is how each structure-specific bone cell could become labelled uniquely, and deposit its labels as appropriate 'cues' on the outer surface of the cell. These surface cues could then help each uniquely specified cell to find its unique place relative to other uniquely specified cells (see Clowes and Wassermann, 1984 for details of the mechanisms). Such a theory could explain why each brain-specific cell becomes located in its correct place. It explains also, indirectly, why all the SMM(cell)s of the SMM(brain) could become located as precisely as is assumed here, as a result of accurate, and unique, matching between ordinary matter brain-specific cells and corresponding SMM(cell)s of the SMM(brain) (see also Wassermann, 1986).

According to the above interpretation, and in agreement with Sperry's views, a brain is not a random assembly of cells, but a highly specifically organized system of, say, for man, a million million cells, no two of which contain the same combination of cell labels.

If the state of affairs, just discussed, and its underlying assumptions are valid, then this could have major implications for my interpretation of OBEs and other psi-phenomena given in chapter 1. By assumption, the SMM(body) and all its parts, including the SMM(brain), are originally formed against the ordinary matter body as a template, during development and maturation. It follows, for instance, that for each synapse that is specified, there exists a corresponding Shadow Matter synapse of the SMM(brain). The likely existence of the vast number of synapses of the human ordinary matter brain would provide a huge number of specified ordinary matter 'contact regions' for the re-entering SMM(brain) at the end of an OBE. This would enable the equally huge number of Shadow Matter synapses of the re-entering SMM(brain) to locate their optimally matching ordinary matter partners. This, in turn, could accomplish highly accurate re-matching between the re-entering SMM(brain) and the ordinary matter brain, leading to re-attachment of the SMM(brain) to the ordinary matter brain. Similar remarks could apply, with appropriate changes, to the likely vast numbers of other cell-cell contacts that could exist between cells of other organs or other structures. These other cell-cell contacts could then facilitate also proper, and highly accurate, re-alignment between the re-entering SMM(body) and various ordinary matter body parts. This, therefore, could answer the question posed on p. 70.

2.1.2 The Possible Role of Shadow Matter in Pattern Recognition

That central nervous nerve cell junctions (i.e. synapses) play a major role in the normal production of intellectual processes is suggested by the fact that certain synapses seem to be involved in forming memory traces for place-learning in rats (see note 4 p. 67). It would be tempting to assume that synapses are the place where memories are formed, and that the ordinary matter brain stores memories. Yet, there are compelling reasons for believing that memories are not stored by the ordinary matter brain at all (see p. 34 and p. 57) and that memories are, probably, normally, and during OBEs, only stored by the SMM(brain) (and that memories can also be stored by SMM(complem, brain)s, see p. 56). It might be objected that even short-term habituation i.e. a form of learning, in the gill withdrawal reflex of the marine mollusc *Aplysia californica* involves a change of synaptic efficacy and hence, involves synapses, (see Kandel and Schwarz's (1982) review and Goelet *et al* (1986, p. 419)).

Yet, granted that, possibly, most normal learning may involve synaptic changes, this does not imply that these changed synapses are the ultimate physical representatives of memories which have conscious epiphenomena when activated. In any case, OBEs suggest strongly that the SMM(brain) can form memories during OBEs without involving synapses. If so, then the SMM(brain) is also the likely location of formation and storage of memories in normal mental processes when the SMM(brain) is assumed to be bonded to the ordinary matter brain. That the ordinary matter brain can, during normal learning processes, activate the SMM(brain), as well as change its synapses in the process, is consistent with the SMM(brain) being the ultimate region of memory formation and storage.

The likelihood that the SMM(brain), and not the ordinary matter brain, is the seat of memories, the region of pattern recognition and the locus where thoughts and feelings are produced, becomes enhanced when we look at *Gestalt* perception. Psychologists, following Köhler, and many others, understand by a Gestalt a complex configuration of, say, an object, which we experience in perception. The Gestalt that is perceived may differ often considerably from the shape of the retinal images of the object. For example, when we look at a circle, located at the centre of a cinema screen, while we are seated very much to the right of the screen, then the retinal image of that circle may be an oval figure rather than a circle. Nevertheless, what we perceive, with our perceptual machinery, looks very much more like a circle than an oval, so that the perceptual machinery somehow corrects the input created by the retinal images. This type of correction is an example of the genesis of 'perceptual constancies' by perceptual machinery. This machinery tries to represent corrected visual images which correspond often more closely to the viewed object than the retinal images.

One of the central problems of visual and tactile perception is to explain, in terms of perceptual machinery, the coherence of perceived configurations, i.e. the coherence of Gestalten. When we look at a motor car, we can certainly make out its doors, windows, roof, etc. Nevertheless, we perceive the car as a coherent whole object and not as a collection of separate items. That this integration of visually presented items into perceived wholes has normally something to do with the ordinary matter brain, may be far from obvious to everybody. It becomes more obvious to sufferers from migraine, who, during an attack, experience scotomata, i.e. blind regions in their visual field. This may lead to a disintegration of a Gestalt into bits and pieces, so that the sufferers simply cannot see the Gestalt as a whole. The great psychologist Wolfgang Köhler was one of those few scientists who recognized fully that Gestalt coherence poses a central problem for any materialistic theory of Gestalt perception.

In the present theory (as distinct from Wassermann, 1978) the visual fields of both retinas are represented collectively by a large multitude of brain synapses, as far as the ordinary matter brain representation goes. Some of these synapses could be located on the same nerve cell, while others could occupy different localities of a host of different brain nerve cells. So how could this host of disjointed, and widely distributed, synapses give rise, when activated, to the representation of a coherent visual field? The problem is as follows: To start with, things that appear to be coherent in experience often are also coherent in nature (e.g. the surface of a motor car, or the window of a house). Next, the human retinas (and many animal retinas) are organized so that their visual receptor cells (the so-called cones and rods) are very precisely 'wired in,' and ordered in specific fashion. They are ordered so that to neighbouring regions of the retinal image of an object there correspond appropriate neighbouring regions of the retina, each region consisting of many precisely ordered visual receptor cells (on each retina). So, why should such an immensely accurate retinal imaging system have evolved, and apparently led to highly accurate, corresponding, perceptual representations (in consciousness)? How could this be if there were not a material ordering system associated with the ordinary matter brain, which leads from the coherent accurate retinal representations to equally coherent and accurate perceptual representations? This suggests that the apparently widely distributed and disjointed collection of synapses that represent the retinas in the ordinary matter brain are unlikely to be the ultimate physical representatives which, when activated, lead to conscious experiences with the correct spatial coherence.

I suggest, therefore, in keeping with section 1.4 (see p. 34) that, normally the SMM(brain) could play also a central role in producing perceptual Gestalt coherence, in association with the ordinary matter brain, although the ordinary matter brain plays no ultimate part in representing this coherence.

Let me elaborate these ideas. I have suggested already, in my interpretation of psi-phenomena, that Shadow Matter is highly elastic and easily expansible and contractible (pp. 45, 46, 49, 50, & 56). Accordingly, the Shadow Matter of the SMM(brain) that is bonded to a particular synapse of the ordinary matter brain could become elastically considerably expanded. It could become extended to such an extent that it can link up, via Shadow Matter bonds, with the Shadow Matter bonded to the nearest-neighbour synapses of the ordinary matter brain. In this way the Shadow Matter representatives of, say, all synapses that are involved in visual image perception could become linked up into a coherent structure which I shall call a perceptual Shadow Matter structure of the SMM(brain). For each sensory modality (i.e. vision, hearing, touch, etc) there could exist a separate perceptual Shadow Matter structure. Within the perceptual Shadow Matter structure each synapse involved in image perception of the sensory modality concerned, is assumed to be represented by an elastically expanded region of the perceptual Shadow Matter structure. Whereas the nearest neighbour synapses of the ordinary matter brain are separated (even if nearby); on the perceptual Shadow Matter structure the regions which represent any two nearest neighbour synapses in Shadow Matter form are assumed to be linked up. Hence, a perceptual Shadow Matter structure of a SMM(brain) represents a continuum of joined up regions, such that each region represents a different synapse. Thus, whereas there exist general discontinuities, in the forms of small or large non-synaptic regions between synapses of the ordinary matter brain, the vast numbers of synapse-representing Shadow Matter structures have no gaps between their nearest neighbour regions, but link up into a single continuous perceptual Shadow Matter structure.

The dynamics of perceptual processes of the kind studied by, say, Gestalt psychologists (see Vernon, 1952), could then be enacted partly by elastic dynamic deformations of the perceptual Shadow Matter structure. I believe that this model does not run into the difficulties encountered by Köhler's model.[2] For instance, for binocular representations there exist nerve cells of the cerebral cortex that respond to corresponding receptive fields of both retinas (as was discovered by Hubel and Wiesel, see Bishop (1970)). Hence, a single minute region of the perceptual Shadow Matter structure, namely a region bonded to a single output synapse of a binocularly driven central nervous nerve cell, could be activated by corresponding receptive fields of both retinas. The activation of the minute region could come about by gravitons emitted by the synapse, i.e. gravitons which enter the perceptual Shadow Matter structure.

To sum up, the preceding application of my SMM(brain) theory suggests that Gestalten are generated normally and in OBEs by perceptual Shadow Matter structures of the SMM(brain). The ordinary matter brain serves only in normal perception to map visual (or sound etc.) images from the sensory receptor surfaces (such as retinas) on to appropriate perceptual Shadow Matter structures of the SMM(brain). In the SMM(brain) Gestalten are then generated as coherent wholes in Shadow Matter form. The wholeness of Gestalten, as distinct from the apartness of ordinary matter brain synapses, is achieved by the postulated continuity and (partial or complete) coherence of most or all parts of the perceptual Shadow Matter structures. Apart from Shadow Matter structures that represent Gestalten, I envisage that there are also Shadow Matter structures that represent memories (engrams) of Gestalten, and Shadow Matter regions that represent concepts.

One brief aside before the end of this section. Professional cognitive psychologists will notice that I have so far completely disregarded the 'feature detecting paradigm' favoured by certain psychologists. According to this, brains recognize patterns by means of nerve cells that can act as feature detectors, detecting such 'features' as corners, edges, curves, and more complex features in hierarchical order. A mere collection of features, however, does not amount to a coherent Gestalt that is perceived (see Rock (1974) and Wassermann (1978) for critiques of the feature-detecting paradigm). Not surprisingly, feature-detecting theories of perception apparently cannot, even in principle, explain how the perceptual machinery could construct Gestalten, their coherence, their perceived depth, colouring etc. (as surveyed by Vernon, 1952, 1962, 1970). Likewise, the feature-detecting paradigm could not explain how the perceptual machinery could construct visual illusions (see Robinson (1972) for many examples). For these reasons alone I agree with Spoehr and Lemkuhle (1982, p. 12), who remarked that:

> The bulk of the psychophysical evidence, nonetheless, fails to uncover in the human visual system the existence of feature detectors as defined by feature models. Rather, these psychophysical findings are more easily explained in terms of response properties of cortical cells as revealed by neurophysiologists. At this time neither the electrophysiological nor the psychological evidence supports the proposition that feature detectors exist and form the basis for all pattern recognition in human beings.

2.1.3 'Stimulus Equivalence' in Everyday Perception and in Out of the Body Experiences

All normal people exhibit the capacity for stimulus equivalence. Lord Brain (1951, p. 23) illustrated vividly the human ability to perceive a wide range of variants of things, seen, heard or touched, as equivalent. He wrote:

When someone speaks and another person listens it may seem a very simple process, but it is so extremely complicated that it is very difficult to understand. Let us take as an example the word 'dog'. No two people pronounce the word 'dog' in exactly the same way. Yet we all know what it means. Not only that, it can be sung, shouted or whispered and it still conveys the same thing. . . But we can go further than that. The written word 'dog' will still mean the same thing, though the word no longer consists of sounds but is made up of black marks on white paper; it can still be understood whether it is written or printed, in large or small letters in black or coloured type, and in any sort of handwriting short of complete illegibility: in many handwritings the marks that people make on paper have little resemblance to the letters they are supposed to represent. Here, then there is as great a variety among the visual patterns presented to the nervous system as among spoken words.

To Brain's illustrations of stimulus equivalence one could add that the word 'dog' can still be recognized in mirror-image writing, or in a book held upside down, though perhaps with difficulty. Lord Brain mentioned also that the sense of touch presents similar generalizing capacities. The reader of braille can recognize a series of pimples on paper which make a pattern quite unlike ordinary letters, and yet also mean the word 'dog'. Stimulus equivalence also exists for many animals (e.g. rats, see Lashley (1960)). If the human perceptual system did not have the capacity to recognize and classify equivalent variants of a stimulus pattern as being equivalent, then we could not recognize most, if any, objects. The retinal image of objects may vary in position on the retina, vary in shapes projected by the objects on the retina, vary in retinal size, colour, etc. and, hence, give rise to appropriate variations in the activated patterns of nerve cells within the ordinary matter brain. Yet, despite this, these greatly varying nervous activity patterns can lead to the correct allocation of equivalent patterns to the same concept (or class).

It can be seen that stimulus equivalence could permit a vast saving in the number of memory traces that a perceptual system has to form. Myriads of variants of a pattern on a sensory receptor surface, say a retina, could have their SMM(brain) representations matched by the same memory trace, provided this memory trace is reversibly deformable. Since by previous assumption all Shadow Matter, and, in particular, all parts of the SMM(brain) are elastically deformable (i.e. highly elastic), this could apply, in particular, to the Shadow Matter representatives of memory traces. This, in turn, could furnish the basis, or part of the basis, of stimulus equivalence.

According to my earlier theorizing in this book, during an out of the body experience part or the whole of the SMM(brain) becomes detached from the ordinary matter brain (together with part or the whole of the SMM(body) of which the SMM(brain) is a part). Then during its out of the body residence the SMM(brain) is the seat of memory traces. It is also the site for the formation of additional memory traces, both during OBEs and in normal perception. Moreover, the assumed SMM(brain) could produce stimulus equivalence during OBEs by the same mechanisms as are used in normal perception for producing stimulus equivalence. Without producing stimulus equivalence during an OBE, the SMM(brain) could not recognize innumerable variants of environmental patterns.

Let me repeat, partially, at this stage some of the major conclusions already reached. The nervous system may be regarded as a huge pre-processing system, which, among other things, channels peripherally obtained information by means of nerve impulses to appropriate central nervous synapses. Via gravitons, these synapses can then pass this information on to SMM(brain) structures which generate what is perceived, thought, and remembered. Perceived colours and perceived three-dimensionality are centrally constructed (see Gibson, 1950; Julesz, 1971, 1975) even if 'cues' for depth perception etc., are peripherally supplied. Since depth perception and colour perception persist normally in out of the body experiences this is consistent with the view that the SMM(brain) is the system which generates the physical representations of perceived depth and colour, and, as already, assumed, also provides the physical basis for a host of cognitive constructs in normal cognition and in OBEs.

2.2 Shadow Matter as a putative Psi-Communicator

2.2.1 A general Mechanism for Psi-Communication

I assumed above (p. 56) that any ordinary matter object whatsoever could form a corresponding SMM(object) which, in most cases, remains firmly attached to that object. I assumed also that a SMM(object) or a part of it, can act as a template for the formation of a SMM(complem) (p. 56), which is complementary to that SMM(object). It was assumed, further (see p.56), that after formation SMM(complem) becomes ejected from its SMM(object) template, to make room for formation of another SMM(complem), and so forth. This assumed process can be symbolized so that we obtain schemata (2.1) and (2.2) namely:

$$\text{SMM(object)} + \text{free Shadow Matter} \rightarrow \text{SMM(object) bound to SMM(complem)} \qquad (2.1)$$

$$\text{SMM(object) bound to SMM(complem)} \rightarrow \text{SMM(object)} + \text{free SMM(complem)} \qquad (2.2)$$

SMM(complem), when ejected from its SMM(object) template, could move off freely into space, and act as a primary psi-communicator. The preceding assumption, like the earlier assumption of SMM(object) formation by an ordinary matter object (pp. 12-3), assumes that there exists always a sufficient concentration of free Shadow Matter of the right kind in space, where required. I assume also as a *central doctrine of my theory* that whereas SMM(object) can generate a SMM(complem), the reverse process is not possible, i.e. a SMM(complem) cannot act as a template for the production of a SMM(object).[3]

In the templating model, just described, a SMM(object) acts, during assembly and bonding of free Shadow Matter as a template in a way analogous to a giant enzyme. It serves to bring together appropriate free Shadow Matter constituents, which by forming Shadow Matter bonds with each other can then form SMM(complem). It follows, from the *central doctrine* stated above, that just as each ordinary matter brain exists only in a single copy, so each SMM(brain) exists only in one copy, possibly in the universe (see note 3). By contrast, SMM(complem)s of parts or the whole of that

SMM(brain) could be formed in vast numbers (see below). I assume also that when a SMM(complem) leaves its SMM(object) template, then the SMM(complem) contracts elastically, and thereby loses what physicists call potential energy which, assuming conservation of energy, becomes converted into kinetic energy of propulsion, thereby leading to rapid ejection of the SMM(complem) away from its template.

Each SMM(complem) is assumed to contain in terms of its Shadow Matter constituents a pattern complementary to its SMM(object) template, although the SMM(complem) differs in Shadow Matter constitution from the Shadow Matter constitution of its template. (The situation is, again, analogous to the relation between a messenger RNA and its DNA template in molecular biology. Although the messenger RNA encodes the genetic message of its DNA template, it does so in terms of ordinary matter units (the so-called RNA nucleotides) which differ from the DNA building blocks (the so-called DNA nucleotides). Thus, the relation of SMM(complem) to its parent SMM(object) template is analogous to the relation of a messenger RNA to its parent DNA template in molecular biology. A SMM(object) is assumed to model the whole, or part, of the ordinary matter object to which that SMM(object) is bound, and model it accurately in terms of Shadow Matter. It follows that a SMM(complem) formed by that SMM(object) could, in its own structural terms, model also accurately, via the SMM(object), that part of the ordinary matter object which binds the SMM(object) concerned. The SMM(complem) can act, therefore, as a carrier of psi-information. It could encode, in terms of its own structure, precise information about the structure of the whole or part of an object, as the case may be. Whether that object is animate or inanimate does not matter, as far as the working of the theory is concerned.

What happens when, during ejection from the SMM(object) template the SMM(complem) contracts, as was assumed above? I stated elsewhere (Wassermann 1988) that:

> Just as an inflated toy balloon with a printed-on pattern, when its air outlet is suddenly opened, starts to contract (losing potential energy) while gaining kinetic energy, a contracting SMM(complem) could remain a coherent entity, retaining the topology of its internal and surface structure, and, thus, the topology of the objects (in the ordinary matter world) which that SMM(complem) represents, in the same way that the toy balloon, while shrinking, retains the topological order of its printed-on pattern.

In the present context 'retention of the topology of a structure' means essentially that although the structure in this case becomes diminished, every part of it, though smaller, remains connected to the same (also smaller) corresponding parts as it was before contraction of the structure occurred.

I make now a further important assumption for my proposed psi-communication mechanism. Let me first explain the motivation for this further assumption.[4] The amplification of copy numbers reached by copying a single SMM(object) into many copies of SMM(complem) would not be sufficient to counter one of the classic objections, which many critics, past and present, have raised against any conceivable physical theory of psi-phenomena. As will be seen (e.g. in the Wilmot case, **Case 15** below), an apparently very intricately structured psi-communication occurred reciprocally between Mrs Wilmot and two other people (one being her husband), over a distance of more than a thousand miles. There are many other spontaneous cases on record in the psi-literature, which involve sizeable distances traversed by psi-communications. (I shall cite some illustrative case histories in the following subsection).

Suppose now, that I had assumed that SMM(object)s and SMM(complem)s are the only SMMs that operate in psi-communications. Then I would possibly be faced by the often cited inverse square law difficulty. I shall now proceed to explain this difficulty for readers who know a little physics. Those readers who know no physics can omit the following paragraph.

Consider an idealized 'point source' which emits photons (i.e. light particles), randomly, in all directions and always at the same rate. Suppose we study the number of photons incident on a square-shaped screen of area one square metre, assuming that the screen is perpendicular to the line joining its centre to the point source. Then the number of photons emitted by the point source per second and hitting the screen will decrease inversely with the square of the distance of the centre of the screen from the point source. Hence, close to the point source there would be a much larger density of emitted photons than further away from that source. In fact, methods for detecting hidden radio transmitters, used by secret agents during World War II, and later, depended on the fact that photon density increases with approach to the source of photon emission. Since no decrease in the quality of psi-communication with distance from the putative 'sender' has been reported, this has been regarded as consistent with the view that photons are not likely transmitters of psi-information (*pace* Upton Sinclair's (1930) radio communication model for psi-phenomena). More generally, the inverse square law difficulty has been taken as an argument against particulate carriers of psi-information, irrespective of the nature of these carriers, and photons were just one example of such carriers.

The present theory can, I believe, overcome the inverse square law difficulty by assuming that there exists a powerful mechanism for the amplification of SMM(complem) numbers, based on indirect replication of SMM(complem)s even while these SMM(complem)s are moving about in space.[5]

I assume also that the mechanism of SMM(complem) amplification could apply even to SMM(complem)s that have become trapped by SMM(object)s that are bound to objects (e.g. buildings). I shall now formulate the amplification mechanism for SMM(complem) numbers. I assume that any SMM(complem), whether freely moving or bound to other stationary Shadow Matter, can act as a template for making a third type of SMM, called simply SMM(third), which is structurally complementary to SMM(complem). It is assumed also that any SMM(third) can, in turn, act as a template for generating other SMM(complem)s, each of the same kind as the SMM(complem) that acted as a template for making that SMM(third).

One can schematize this SMM(complem) amplification process as follows by means of schemata (2.3) and (2.4):

SMM(complem) + free Shadow Matter $\rightarrow$ SMM(complem) + SMM(third)   (2.3)

SMM(third) + free Shadow Matter $\rightarrow$ SMM (third) + SMM(complem)   (2.4)

Where it is assumed that SMM(third) in (2.3) leaves its SMM(complem) template immediately after formation, and *mutatis mutandis* SMM(complem) in reaction (2.4). In the presence of the assumed rich supply of free Shadow Matter (in the universe, see p. 77) there could, by this reciprocal templating mechanism, build up a large supply of SMM(complem)s and complementary SMM(third)s of appropriate kinds. One cannot rule out this reciprocal templating mechanism on preconceived grounds, since a closely analogous mechanism has evolved in the world of ordinary matter, namely the process of DNA replication (see note 5). The reciprocal templating mechanism could overcome the inverse square law difficulty since it can generate afresh a rich supply of SMM(complem) copies even at a distance very far removed from the original SMM(object) source that generated the particular type of SMM(complem). One does not have to assume that SMM(complem) and SMM(third) remain stationary while acting as templates, any more than one has to assume that an enzyme molecule has to remain stationary while acting on other molecules (so-called substrates). Also, just as there exist reaction assisting enzymes in the world of ordinary matter, which greatly speed up reactions, so there could exist reaction assisting entities in the Shadow Matter world. Such entities could assist reactions of type (2.1), (2.2), (2.3) and (2.4) among others.

If amplification of SMM(complem) and SMM(third) numbers, as assumed in schemes (2.3) and (2.4) for any particular type of SMM(complem), were the exclusive mechanism for the control of SMM(complem) and SMM(third) copy numbers, then this would lead to serious difficulties. For each particular existing type of SMM(complem) there could occur an indefinite amplification of the number of SMM(complem) and SMM(third) copies, with concomitant decrease in the amount of available free Shadow Matter. This could be avoided by a Shadow Matter mechanism which ensures that numerous SMM(complem) copies and SMM(third) copies of each type become broken down also in sufficient numbers. This could ensure that there exists a steady state for any existing type of SMM(complem) and related SMM(third) type which exists on Earth. In this steady state the rate of further SMM(complem) and SMM(third) production is, on average, balanced by the rate of destruction, by breakdown, of these particular SMM(complem) and SMM(third) types. The control mechanism could operate by negative feedback, so that the larger the number of SMM(complem) copies for any one type of SMM(complem) in any locality, the higher the rate of destruction of SMM(complem) of that type in that locality. (Similar remarks could apply to the SMM(third) copies of the associated type in that locality.) It can be seen that this postulated mechanism is in some respects analogous to the 'turnover' of specific molecules and macromolecules in the case of ordinary matter structures of organisms.

Molecular turnover in organisms requires, as a rule, a large number of types of macromolecules (enzymes) for the total turnover of all molecules involved (anabolism and catabolism, as it is technically known). This raises the thorny question of how all the 'right' molecules that control the turnover of various macromolecules (enzymes) could have evolved with their correct adaptive features.

To cope with this question I suggested a detailed mechanism how such macromolecular adaptations could occur non-randomly, and called this mechanism TIMA (= template induced molecular adaptation) (Wassermann, 1982b, 1982c). In an analogous way Shadow Matter structures, corresponding to ordinary matter molecules, could have evolved and participate specifically in the turnover of SMM(complem)s and SMM(third)s in highly selective ways.

2.2.2 Evidence that psi-communications can occur over very long distances

Since SMM(complem) amplification, and its linked SMM(third) amplification could permit psi-communication over very long distances without signal attenuation (i.e. loss of spatial density of SMM(complem) numbers etc.), I wish to cite a few cases that seem to involve long-distance psi-communication.

Case 8

This case was reported by Myers (1903, vol.2, p. 395) and was taken by him from Gurney *et al.*(1886) (case 146). Myers summarized the case as follows:

It concerns 'The well known incident recorded by Lord Brougham - his vision, while taking a warm bath in Sweden, of a school friend from whom he had parted many years before, but with whom he had long ago 'committed the folly of drawing up an agreement written out with blood, to the effect whichever of us died first should appear to the other, and thus solve any doubts we had entertained of the life after death.' This incident happened about 1am apparently on December 19th (possibly December 20th) 1799. G died in India on December 19th 1799 - place and hour not stated. The time in any part of India is, of course several hours ahead of the time in Sweden.'

The putative psi-communication in this case involved the distance between Sweden and India. The friend's death could have involved complete severing of his SMM(body) (including his SMM(brain) from his dead ordinary matter body, and survival of that SMM(body) and SMM(brain). The surviving SMM(body) could then have generated sequentially many SMM(complem, body)s which also had SMM(complem, brain)s which could encode the information also carried by SMM(brain) of G. The SMM(complem, body) of G, could then have travelled randomly, while amplification of numbers of that SMM(complem, body) took place. One of the copies made of some descendent of the original SMM(complem, body) of G, could then have reached the SMM(body) (including the SMM(brain)) of Lord Brougham, which then activated, via gravitons, the ordinary matter brain of Lord B and gave rise to speech that represented verbally some of the information transmitted by the SMM(complem, body) of G.

This interpretation is not only consistent with my theory of psi-communication, but also with the possibility of the survival of the SMM(body) and its SMM(brain) of Brougham's friend G. The precise manner of interaction between Lord Brougham's SMM(body) (including his SMM(brain)) and the SMM(complem, body) of G, which led to Lord Brougham's recognition of G would demand much further specification of the mechanism, with many additional ad hoc hypotheses, than occurs in my theory. I therefore leave the case, noting that my theory permits us to understand how information was carried from the dead friend G to Lord Brougham in Sweden.

Just as in ordinary perception so in psychic perception (i.e. ESP) there must exist attention mechanisms. These could be the same as those for normal sensory perception and could be mediated by the SMM(brain).

Case 9

Next, consider the case of an apparition of a living person, who, while in a crisis in England (in 1864) appeared at that time as an apparition simultaneously to two percipients in Cairo. Only one of the percipients knew the man, but the other gave an accurate description of what she had witnessed. The case is reported by Myers (1903, vol.1, pp. 287-9) and taken by him from Gurney *et al.* (1886, vol.2, p. 239). The case, like **Case 1**, also involves several percipients, showing that simultaneous apparitions, such as in Case 1, form a class with repeatable class characteristics, namely that the apparition occurred simultaneously to two or more people (which was not so in **Case 8**). Again in the present case, psi-communication seems to have operated over a long distance. Hence, **Cases 8** and **9** also share a class characteristic, as far as long-distance psi-communication is concerned. Yet, one of the two percipients in Cairo could have perceived the psi-information from the other by telepathy (see **Case 1** p. 24, and for a possible mechanism of telepathy see below).

In the case, just discussed, the SMM(body) of the living person, at the time of crisis, could have become more weakly bonded to the ordinary matter body than usual, and given rise to a sequence of SMM(complem)s which, by means of SMM(complem) amplification reached the SMM(body) of at least one of the two percipients in Cairo.

Case 10

This case, reported by Myers (1903, vol.1, p. 687), was taken from Gurney *et al.* (1886, vol.2, p. 227) and supplied by Commander T. W. Aylesbury (late Indian Navy) of Sutton Surrey. Like the Wilmot case, reported later in this book, the present case seems to involve a reciprocal apparition which was also a crisis apparition. It occurred, putatively, by psi-communication over a very long distance and was reported in December 1882. The case could be explained in terms of psi-mechanisms just like the preceding case. Myers reports that:

The writer (Commander Aylesbury), when thirteen years of age, was capsized in a boat, when landing on the island of Bally, east of Java, and was nearly drowned. On coming to the surface, after being repeatedly submerged, the boy called his mother. This amused the boat's crew, who spoke of it afterwards, and jeered him a good deal about it. Months after, on arrival in England, the boy went to his home, and while telling his mother of his narrow escape, he said, 'While I was under water I saw you all sitting in this room; you were working something white. I saw you all mother, Eliza and Ellen.' His mother at once said, 'Why, yes, and I heard you cry out for me, and I sent Emily to look out of the window, for I remarked that something had happened to that poor boy.' The time owing to the difference of English longitude corresponded with the time when the voice was heard.

Myers added the testimony of the event written by one of Commander Aylesbury's sisters. Note that the geographical separation was between Java and England. A full explanation of this case could be given in terms similar to those of **Case 15**. For instance, it could be assumed that during his crisis, when under water, the boy's SMM(brain) became more weakly linked to his ordinary matter brain. This could then have given the boy access to clairvoyant signals in the forms of SMM(complem, object)s coming from his mother's home. (see also **Case 9** for a similar explanation; and see my theory section 1.6, particularly pp. 38-9)

Case 11

Another interesting crisis apparition which involves a large separation between the percipient and the person involved in the crisis (or death in this case) was reported in the *Proceedings of the Society for Psychical Research* (London) vol.33, p. 170 and was summarized by Tyrrell (1953, p. 36) as follows:

> The percipient's half-brother (she refers to him as her brother), an airman, had been shot down in *France* on the 19th of March 1917 early in the morning. She herself was in *India.* 'My brother,' she says, 'appeared to me on the 19th of March 1917. At the time I was either sewing or talking to my baby - I cannot remember quite what I was doing at that moment. The baby was on the bed. I had a very strong feeling that I must turn round; on doing so I saw my brother, Eldred W. Bowyer-Bower. Thinking he was alive and had been sent out to India, I was simply delighted to see him, and turned round quickly to put baby in a safe place on the bed, so that I could go on talking to my brother; then turned again and put my hand out to him, when I found he was not there. I thought he was only joking, so I called him and looked everywhere I could think of looking. It was only when I could not find him I became very frightened and the awful fear that he might be dead. I felt very sick and giddy. I think it was two o'clock the baby was christened and in the church I felt he was there, but I could not see him. Two weeks later I saw in the paper he was missing. Yet I could not bring myself to believe he had passed away.' (Italics are mine.)

After death of the airman his assumed surviving SMM(body) could have separated from his dead ordinary matter body and could have generated and emitted SMM(complem)s. The latter, by the earlier proposed amplification mechanism, (see (2.3) and (2.4) p. 80) could then have generated many copies of these SMM(complem)s, some of which reached the percipient's SMM(brain) which then can synthesize the apparition. As explained subsequently in my theory of clairvoyance and telepathy, the amplification mechanism ensures that vast numbers of copies of an emitted SMM(complem) could be generated, all of which make random journeys, and only one or a few of them reach the SMM(brain) of the percipient. This is the manner in which contact between emitter and percipient is established, according to the theory in the present case. (See also comments below). That people can synthesize such complex apparitions is suggested by the post-hypnotic apparition which was suggested to a subject by Gindes (**Case 3a**, p. 42).

In **cases 8-11** I have chosen deliberately those involving large distances covered by the apparent psi-communication (see also **Case 15** for another such example). Numerous cases of similar types exist, involving only a few miles between percipient and event (e.g. some travelling clairvoyance cases; see Tyrrell, 1953, pp. 119-24, or alternatively *Proceedings of the Society for Psychical Research* (London) vol.7, p. 206, vol.41, p. 345, vol.7, p. 58).

2.3 A Proposed Shadow Matter Mechanism for Telepathy

I suggest that telepathy works as follows. To start with the SMM(brain) of person S, who is the telepathic sender, becomes, partly or completely, temporarily weakly bound to the ordinary matter brain of person S (while person S is alive). In more extreme cases the SMM(brain) of person S becomes partly or completely detached from the ordinary matter brain, either during an 'out of the body experience' or following death. When in this weakly bound or detached state, then the SMM(brain) can generate numerous SMM(complem)s which could each represent Gestalten currently actively represented by the SMM(brain). Or the generated SMM(complem)s could represent memory traces stored by that person's SMM(brain) or thoughts going on in that SMM(brain) etc. Following the genesis of SMM(complem)s, of the kinds mentioned, these SMM(complem)s could be ejected from the living or possibly surviving (in **Case 11**) SMM(brain). The ejected SMM(Complem)s while moving in space, could replicate by the proposed amplification mechanism (p. 80). This could lead to the production of a large number of copies of each of the ejected SMM(complem)s. One or more of these copies of SMM(complem)s, while freely travelling, could be captured by the deciphering, matching, machinery of SMM(brain) of the telepathic receiver R whose SMM(brain) is weakly bound to, or detached from, his or her ordinary matter brain. The cognitive contents of the captured SMM(complem) could be recognized via pattern matching by appropriate memory traces of the SMM(brain) of the receiver, by making use of the normal sensory perception machinery etc., of that SMM(brain), provided the required memory traces, etc., exist. The theory assumes that the SMM(brain)s of the telepathic sender S and receiver R represent any particular thought, memory, etc., by means of the same or similar material code.

The mechanism, just formulated, could apply also to telepathic transmission of the contents of concept-representing memory traces, such that such a memory trace of sender S, even if he is not consciously aware of its activation at the time, could lead to activation of a corresponding memory trace of receiver R. This theory of 'telepathic memory communication,' which is a subordinate part of my general theory of telepathy, could explain, in principle, how any telepathic receiver R could have access to all memories of any other person on Earth, provided the SMM(brain)s involved are, at least temporarily, sufficiently weakly bound to the ordinary matter brains. Although there exists much circumstantial evidence for such 'super-telepathy,' based on numerous case histories, many people have doubted it, because they could not envisage any conceivable mechanism. At least one such mechanism has now been provided within an integrated theory of psi-phenomena.

Let me now summarize the gist of my theory of telepathy in the following schematic representation.

Schematic Representation of the Proposed Mechanism of Telepathy

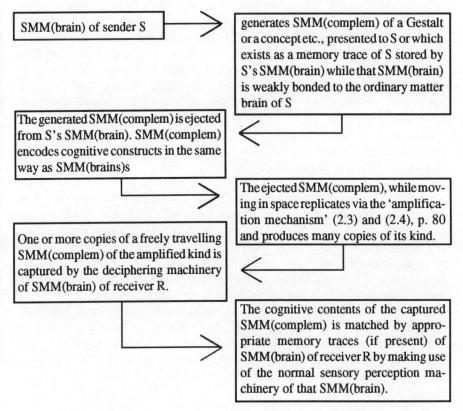

2.4 A Proposed Shadow Matter Mechanism for Clairvoyance

According to my theory any ordinary matter object normally binds a SMM(object). There exist, within the present theory, two possibilities. Under suitable conditions (e.g. when sunlight is reflected) an ordinary matter object could emit photons and also sphotons (= Shadow Matter photons; (see p. 36) that are bound to the photons. Alternatively, or in addition, the ordinary matter object's attached SMM(object) could, in suitable conditions, emit sphotons, say, from its surface. Let me now suggest a major mechanism for clairvoyance which could account also for 'travelling clairvoyance' (as illustrated in chapter 3). This mechanism assumes that the SMM(body) of any human being P (including the SMM(brain)) can emit completely complementary SMM(complem, P)s which are complementary to the set of ingredients of SMM(body). Thus, to use an inadequate analogy, SMM(complem, P) of person P compares to SMM(body) of P like a photographic positive to a photographic negative. It is also assumed that SMM(body) of person P can form and release, in rapid succession a large number of SMM(complem, P)s, which after release, move off into

85

space and replicate (by the amplification mechanism of (2.3) and (2.4), p. 80). The replicas, thus formed, may, in turn, replicate, as stated in connection with the amplification mechanism. (The difference between the complete SMM(complem, P) release from the SMM(body) of person P, and the incomplete release of SMM(body) from an ordinary matter body in an out of the body experience (OBE), is, according to my theory, due to the circumstance that the SMM(body) remains attached to the ordinary matter body by a Shadow Matter cord during the OBE (see p. 50), whereas the SMM(complem, P) when emitted from the SMM(body) of person P, becomes completely separated.)

Just as in an OBE the SMM(body) of person P can, by assumption, perceive its environment with the help of its Shadow Matter eyes and its SMM(brain), when the Shadow Matter eyes receive sphotons (see p. 36-7 emitted by environmental objects, so it is assumed that in clairvoyance any SMM(complem, P) can, via its Shadow Matter eyes receive sphotons derived via photon-sphoton pairs (p. 37) emitted from environmental objects. (The Shadow Matter eyes of SMM(complem, P) are assumed to be complementary to the Shadow Matter eyes of the SMM(body) of person P.) The preceding events could then cause SMM(brain) of SMM(complem, P) to perceive the environmental objects (but not necessarily consciously, since the SMM(brain) of SMM(complem, P) need not give rise to conscious experiences). Also, in this kind of theory it need not be assumed, at least in its present formulation, in which way the SMM(brain) of SMM(complem, P) represents perceived external objects. There is, for instance, no need to assume that the representation is isomorphic, i.e. that the representation resembles in form the external object which is being represented. The representation could be in a totally different code.

In this type of theory the information receiver, i.e. SMM(complem, P) or one of its replicas (by the amplification process (2.3) & (2.4) p. 80) could then, by travelling randomly through space, get within range of sphotons emitted as sphoton-photon pairs (p. 37) by one or more objects. These objects could be far removed from person P. When this happens the sphotons could activate the Shadow Matter eyes of SMM(complem, P), leading, in turn, to the activation of SMM(brain) of SMM(complem, P). This could lead to the formation of Shadow Matter memory traces by SMM(complem, P). These memory traces could then represent the objects encountered by SMM(complem, P) during its free travels. Subsequently, when either this SMM(complem, P) or any of its descendants via the amplification process (2.3) and (2.4) happens to make contact with SMM(body) of person P, during random movement of SMM(complem, P) the following could happen. The SMM(complem, P) and the SMM(body) of P could interact. Then the SMM(brain) of person P could form additional engrams which correspond to those engrams formed of the objects encountered by that SMM(complem, P) or one or more of its ancestors, during travel.

This, then, is the basis of the present theory of clairvoyance, in which emitted SMM(complem, P)s, and their descendants by amplification, scout the environment, and some of these descendants by chance return ultimately to the source of emission of the original SMM(complem, P), namely person P.

The following schematic representation, given below (p. 87) provides the gist of my theory of clairvoyance, as described and explained above.

I must recall that SMM(complem, P), when encountering the SMM(brain) of person P, can only interact (via SMM(brain) of SMM(complem, P)) with SMM(brain) of P provided the latter is temporarily weakly bound (at least in certain regions) to the ordinary matter brain of person P. (See pp. 38-9 above). For instance, in **Case 17** below, the SMM(brain) of the subject could have separated from the ordinary matter brain during the OBE phase of the phenomenon, which then facilitated travelling clairvoyance.

Schematic Representation of the Postulated Mechanism of Clairvoyance[6]

| SMM(body) of person P | → | emits SMM(complem, P) which, among other things, contains complementary copies of SMM(brain) and SMM(eye)s of person P |

SMM(complem, P) triggers an amplification process of type (2.3) and (2.4) p. 80, which yields many replicas of SMM(complem, P). These replicas, while replicating further, move about randomly. While doing so they encounter all kinds of environmental objects, including people. These objects emit Shadow Matter photons (=sphotons) as sphoton-photon pairs (see p. 36-7)

Sphotons, thus emitted, interact with Shadow Matter eyes of SMM(complem, P) copies, and, via these Shadow Matter eyes interact with SMM(brain) of SMM(complem, P) copies, where memory traces of encountered objects may be formed

SMM(complem, P) copies and their further replicas can continue to move about randomly and one or more SMM(complem, P)s could then encounter SMM(body) of person P and interact with it

Thereby SMM(brain) of person P could form additional memory traces corresponding to those formed by SMM(complem, P) or its ancestors while they encountered objects in the environment. SMM(complem, P)s, thus, serve as information gatherers about the environment, which are emitted by SMM(body) of person P and some of whose descendants, arising from the amplification process ultimately return to the SMM(body) of person P

As stated, the SMM(brain) of person P could acquire additional memory traces by encounters with SMM(complem, P). These extra memory traces could correspond to distant objects or situations. They can then be used by that SMM(brain) of P in order to construct spatially and serially ordered, clairvoyantly experienced, hallucinations of distant things, people, scenery, etc. The process of construction could lead often to quite vivid visual, etc., perception. That process could rely on the SMM(brain) machinery that is, by assumption, involved in the hallucination reported by Gindes (**Case 3a**, p. 42). In that hallucination a young woman perceived post-hypnotically her dead brother, who showed perfectly normal behaviour, normally integrated into the environment. Various cases of clairvoyance and travelling clairvoyance could be explained along the lines of the general mechanism of clairvoyance and will be cited in some detail below (**Cases 15-26** and **Case 28**).

Two further points require elucidation and comment as regards the preceding theory of clairvoyance. First I must discuss the problem of object coherence in clairvoyance. A theory of clairvoyance must explain how a large set of objects (e.g. a table top, the table legs, and the carpet on which the legs stand) cohere. For, innumerable sets of objects that, when seen in normal perception, appear as a coherent system of objects, are also experienced as similarly cohering objects when perceived clairvoyantly (see numerous cases cited in the following chapter, typically **Cases 15-16** and **18-20** and others). If, instead of the preceding theory, I had assumed that clairvoyance involves that each object of a set of coherent objects, emits a SMM(complem) specific for that object, then this would not explain how, when the various SMM(complem, object)s reach a human percipient, they are perceived as a coherent system. By contrast, the theory postulated above (pp. 85 ff) assumes that Shadow Matter bound to the objects in the guise of SMM(object)s emits sphotons. These sphotons, when acting on Shadow Matter eyes of a SMM(complem, P) can induce the SMM(brain) of SMM(complem, P) to produce a coherent representation of objects, as a result of images of coherent objects produced by the sphotons on the Shadow Matter eyes. All this is quite analogous to the manner in which photons emitted by coherent ordinary matter objects lead via retinas of ordinary matter eyes to the assumed representation of coherence by SMM(brain)s of people.

The second problem concerns the clairvoyant ability to see clairvoyantly into houses, to become aware of the contents of closed books and sealed envelopes (see various examples in chapters 3 and 4). One simply cannot deduce from the fact that ordinary matter photons of visible light can pass through (sufficiently thin layers of) water and glass (if transparent), but not through optically (non-transparent) walls of houses, that the same need necessarily apply to sphotons. Since Shadow Matter and ordinary matter interact only gravitationally, there are no reasons why sphotons should not penetrate through optically impenetrable ordinary matter.[7]

It can be seen that my theory is unificatory in that it explains telepathy and clairvoyance in terms of essentially the same types of Shadow Matter mechanisms (involving propagation of SMM(complem)s). In fact, it is possible that the mechanism for telepathy could resemble that of clairvoyance, described above, even more closely than the mechanism suggested in section 2.3. It was assumed in section 2.3 that only SMM(brain)s emit SMM(complem)s in telepathy. One could have, instead, that

SMM(body, P) of person P forms and emits SMM(complem, P) which then replicates via the amplification mechanism of schemata (2.3) and (2.4) (p. 80). When SMM(complem, P) or one of its descendent replicas encounters the SMM(body, R) of some receiver person R, then, in suitable circumstances (notably of not too tight binding to the ordinary matter brain), the SMM(brain) of SMM(complem, P) could interact with the SMM(brain) of person R. In this way one person could telepathically perceive SMM(brain) representations of another.

I must stress also that SMM(complem)s which, by assumption, serve as vehicles for psi-communications, and for gathering of psi-information from the environment, both in telepathy and clairvoyance, are, unlike sphotons, highly structured. It could be expected that the structural complexity of SMM(complem)s that are complementary to SMM(body) of person P are as complex, in terms of Shadow Matter, as the anatomical structure of person P is in terms of ordinary matter. Thus, if the mediators of psi-information are highly complex systems, this could explain why 'radiation field theories,' notably those based on photon transmission, 'cannot cope with the informational aspects of psi-communication' (Beloff, 1979). From this, however, one cannot conclude that no conceivable physical theory could explain psi-communications. The present theory provides one potential counter-example against the views of those who, like Beloff (1970), entertain such thoughts.

Alas, when it comes to psi-phenomena Beloff is not the only one who thinks that he knows how the phenomena could not work (see p. 16). At the more extreme end is James Randi. His tactics are simple. Typically he picks out say a single, supposedly 'well known' psychic medium. Then, when the medium is asked to perform for television at Randi's appointed time, that psychic fails to perform adequately (see Randi, 1991). This may leave some, or perhaps many, viewers of the television programme, notably those who do not know any statistics, in doubt whether, just this medium, or all ever existing mediums could perform adequately, and hence, that psychic phenomena are figments of the imagination of parapsychologists. The facts, of course, are different. I argued on p. 39 that really outstanding trance mediumship is rare, and that trance may involve weakened bonding between SMM(brain) and ordinary matter brain. If so, we cannot expect that any 'medium' can always be in a suitable biological state for receiving, say, telepathic communications, or clairvoyant communications (e.g. just whenever demanded by Mr Randi (see 1991)). Not surprisingly, when asked to perform on demand these mediums may, indeed, sometimes, resort to guessing. Yet, from the fact that a single medium or handful of mediums cannot perform 'to order,' when requested by Mr Randi on television, we cannot conclude that very prominent trance mediums, like Mrs Piper and others (see p. 39) have not demonstrated extraordinary mediumistic feats, certainly not based on guessing. If outstanding trance mediumship applies only to, say, one in a thousand (or much fewer) supposed mediums, then picking one allegedly 'gifted' medium (who may only be mildly gifted) and asking him or her to display on TV true talent just when it suits you, looks to me like wrong methodology.[8] By looking only at relatively very few mediums on television, and letting them only try a few things, one is restricting oneself to a statistically inadequate sample size. Mr Randi disregards also that most psi-phenomena are spontaneous, and therefore cannot be elicited at any time and in any way that suits him. His failure to provide positive evidence on a TV show, when and how it suits him, proves nothing except that he may not have tried long enough and hard enough.

Let me now return to my psi-mechanisms. Just as the antenna of a radio receiver can be bombarded continually by a host of photons (i.e. particles of electromagnetic radiation fields) of different frequencies, emitted by a large number of different radio transmitters, so the SMM(brain) of every human being (and, presumably, also of many kinds of animals) could be bombarded constantly by SMM(complem)s. The latter could be directly or indirectly derived from SMM(object)s, including SMM(brain)s of other people. Selectivity in psi-perception could be ensured by the 'attention mechanisms' with which the SMM(brain) is assumed to be endowed, and which, according to my theory, also operate in normal perception. Just as in normal sensory perception these attention mechanisms could ensure that the SMM(brain) does not become overwhelmed by environmental information, something similar could happen in extrasensory perception (ESP). In ESP the attention mechanism of the SMM(brain) could ensure that environmentally arriving SMM(complem)s are not all equally responded to even when the SMM(brain) is in a state where it can respond.

2.5 Shadow Matter explanations of certain kinds of apparitions, haunted houses and 'object reading'

I suggested already elsewhere (Wassermann, 1988) that, for instance, the SMM(body)s of some residents of buildings could sometimes give rise to SMM(complem, body)s, which represent their bodies, including the clothes they wear. Some of these generated and emitted SMM(complem, b)s would then get selectively 'captured' by the SMM(object)s bound to that building (e.g. the walls of the building). Only some kinds of structures and objects and materials might capture SMM(complem,b)s (i.e. only some buildings are haunted). Some of these captured SMM(complem, b)s could then, while remaining captured (on a long-term basis), emit sphotons. Some of these sphotons could then impinge on the Shadow Matter eyes of visitors, or new inhabitants, of the same building, and could then generate apparitions via the SMM(brain)s of the visitor. This, according to the theory, could only happen provided the SMM(brain) is, at least in parts, sufficiently weakly bound to the ordinary matter brain, and provided the Shadow Matter eyes are at least temporarily sufficiently weakly bound to the ordinary matter eyes. The SMM(brain)s of the visitors concerned could autonomously construct the apparition in a manner resembling what happened in the post-hypnotically produced apparition described in **Case 3a**. The preceding mechanism could form the physical basis of various types of apparitions associated with haunted houses of the kinds classified by Tyrrell (1953). (See also Green and McCreery (1975) for examples of various kinds of apparitions.)

As an alternative mechanism it could be assumed that SMM(complem, b)s of former inhabitants, which are now bound to the, say, SMM(walls) of the walls of a building, produce by replication, via the mechanisms (2.3) and (2.4) p. 80, many additional copies of SMM(complem, b)s, with the building-bound SMM(complem, b)s acting as the primary templates for replication, in the first instance. Some of these replicated SMM(complem, b)s could then interact with the SMM(brain)s of people visiting or inhabiting the building.

Apparitions of former inhabitants of a building occur notably when percipients are inside the building, or outside but nearby. If the sphoton-based assumed possible mechanism is operative, then, because of an expected sphoton attenuation with distance, one could expect that nearness to the building would favour vision of a SMM(complem, b) by a percipient while near the building.

Some haunted houses are quite modern (see Tyrrell, 1953, p. 38 for a typical case history; or see *Proceedings of the Society for Psychical Research* (London) vol.3, pp. 102 ff). Several of these apparitions (or 'ghosts,' as they are called popularly, if they are apparitions of deceased people) were experienced by several people, either simultaneously or at different times (e.g. in the case just mentioned, and in another case discussed by Tyrrell (1953) pp. 55-6 and also found in the *Proceedings of the Society for Psychical Research* (London) vol.8, pp. 311 ff). The present theory could explain such cases as follows in terms of the first of the two alternative mechanisms given. Several copies of the same type of building-bound SMM(complem, b), representing a former inhabitant of a building (still alive or since deceased), could interact, via sphotons with the Shadow Matter eyes of percepients. This could activate the SMM(brain)s of several percipients either simultaneously or at different times, when these percipients are close enough to the building to which these SMM(complem, b)s are bound. The activated SMM(brain)s could then construct the apparition. Let me cite some examples.

Case 12

Tyrrell (1953, p. 38) cites that a lady was playing cricket in the garden (i.e. near her house) with her little boys. She described that:

From my position at the wickets I could see right into the house through an open door, down a passage through the hall as far as the front door. The kitchen door opened into the passage. I distinctly saw the same face [that of an apparition previously repeatedly experienced by her within the same building] peeping round at me out of the kitchen door. I again only saw the upper half of the figure [as on previous occasions when the same subject experienced the same apparition]. I threw down the bat and ran in. No one was in the kitchen. One servant was out and I found the other was up in her bedroom. . .

Case 13

Summarizing another case Tyrrell (1953, pp. 55; and see also Morton (1892)) writes:

The well-authenticated ghost, studied by Miss Morton, which was seen and heard by a number of people over a period of years, proved its non-physical character in many ways. Miss Morton fixed threads across the stairs, very tightly secured by pellets of marine glue, and twice watched the apparition pass through them without disturbing them.[9] She was also frequently watching the apparition when it disappeared in its accustomed place near the garden door. Towards the end of its existence it faded away gradually.

Tyrrell's remark that the apparition, just mentioned, was 'non-physical' was simply in keeping with the physics of his time, when Shadow Matter was undreamed of, at least in the way introduced above. According to the present theory, apparitions are very physical indeed, although their physics is assumed to belong to the realm of Shadow Matter theory. By this, I do not mean that the apparition, as such, exists where it is experienced, say on the staircase in the case of the Morton ghost. I simply mean that there exist (say on the staircase) physical Shadow Matter representations of the apparition in the forms of SMM(complem, b)s and that the apparition could then be generated by the percipient's SMM(brain) by, say, one or the other of the mechanisms described above.

Another psi-phenomenon, whose mechanism, I believe, is closely linked to my postulated mechanisms for explaining haunted houses, is so-called 'object reading,' also popularly known as 'psychometry'. Since the term psychometry has a different meaning for experimental psychologists, the term object reading will be preferred. Before I present my theory of object reading I shall first cite a few cases.

Object reading has been extensively studied with several sensitives by Dr Eugene Osty (1923), formerly Director of the Institut Métapsychique International, Paris. Likewise Pagenstecher made an extensive study of one sensitive, Señora Z, (*Proceedings of the American Society for Psychical Research* vol.15, 189-314 and vol.16). As Tyrrell (1948, p. 186) put it 'The influence of an object in assisting a sensitive to become aware of facts about its owners or those who have touched it is well brought out by the case of Señora Z,' the daughter of the Governor of the State of Michoacan in Mexico.

Dr Pagenstecher, a physician of high repute, made Señora Z do her object reading under hypnosis.[10] Tyrrell writes

Various objects were given her. Sometimes she would go back to the beginning and describe the scene of their manufacture; sometimes events in which they had played a part. A fragment of marble, for example, taken from a temple in a Roman Forum produced a recognizable description of the Forum as seen from a particular point of view, although she had not been told where the marble came from. Pumice stone, taken from the bed of Lake Texoco, produced not only a description of its volcanic origin (perhaps obvious) but also a description of fishes swimming above it. Some of the information might have been telepathically acquired from. . . Dr Pagenstecher.

Here we see what is typical of most or all object reading. The sensitive who holds the object 'reads out of it' what comes spontaneously his or her way. This spontaneous activity in which the psychic divulges what information the object gives him or her, without being prompted to get specific information, is quite different from what James Randi (1991, pp. 85-6), a 'psychic investigator,' believes most object reading to be. He confronted a psychic in an object reading test with definite identification tasks concerning which objects belonged both to the same owner. In this non-spontaneous task the psychic failed. All this demonstrates that psychic faculties that normally manifest themselves spontaneously need not be able to function 'to order,' a point already made repeatedly, but, unfortunately, not properly recognized by James Randi. The following case illustrates this spontaneity.

Case 14

Consider the following account by Tyrrell (1948, p.180) based on Osty's studies of object reading.

Osty received a letter enclosed by a certain Captain C and was told only that the writer of the letter was now dead. On the 18th May 1922, he gave this letter to Mme Viviana, who crushed it in her hand and said the writer was dead; a soldier; in the war; sunburnt; had a very direct gaze; was strong-willed and combative; unsentimental; intelligent; good; energetic; amiable; Catholic; had a tendency to mysticism; would pray when sad or troubled; not bigoted; high-minded; came of a religious family and from a country where they gave boats the names of saints, as in Brittany; had an elder brother in whom he placed much confidence; his only anxiety was for a dearly-loved woman; there is a child; a feeling of swaying, rolling; of humidity and water... as if he were on the water; 'my lips are salt as if I were on the sea;' an officer; young; died at the end of the war; not from a wound; suffocation; a sudden pain in the head; did not die in bed; small houses, soldiers, engineers, pick-axes, tents round him.

The points of information are here condensed. Certainly the letter could not have given all this information even if it had been read by the sensitive.[11]

Twenty five of these points were found to be correct, and four are unverifiable. There was no false statement. The writer was Captain C's brother and the letter had been written at sea in the transport *St Anne* in rough sea. Hence, the feeling of swaying and rolling and of salt on the lips. The writer had been out at the war on the Balkan front, but was not killed in battle. He was reported to have died of influenza.

Although a good many items in the sensitive's statements could apply to almost anybody, a few are impressive, although some of Osty's best cases that I have read are substantially more impressive.

Let me now interpret object reading within my theory, in terms very similar to my explanations of haunted houses. My theory of haunted houses suggests that the SMM(object)s bound to buildings, and modelling the ordinary matter structures of these buildings in Shadow Matter form, could 'capture' and retain SMM(complem)s generated, directly or indirectly, by other SMM(object)s (including SMM(body)s of people who frequently visit or reside in these buildings). In much the same way, people, or other objects, who were close to an object O, for a sufficiently long time, could have transferred appropriate SMM(complem, transferred)s to the SMM (object) of object O, these SMM(complem, t)s being complementary to SMM(object)s of the objects close to O. The SMM(object)s of object O could then have captured and retained (on a long-term basis) one or more copies of each SMM(complem, t) transferred to it. In this way, successive owners, carriers, or other people touching the object O, or who are simply close to it (without touching it), could have some of their emitted SMM(complem, t)s transferred to, and trapped by, the same object O. For instance, SMM(complem, t) emitted by 'Shadow Matter represented' memory traces (engrams) of a SMM(brain) of a person who carried object O could encode these memory traces in a form trapped by the SMM(object) of object O.

Just as the SMM(complem, b)s, bound to the SMM(object)s of a building, could give rise to emitted copies of these SMM(complem, b)s via the previously assumed replication process of SMM(complem)s (see p. 80), so any SMM(complem, t) bound to a SMM(object) of an object O could, after capture by that SMM(object) give rise to emitted copies of that SMM(complem, t). When object O is handed to, and thus close to, a 'sensitive,' one or more of the emitted SMM(complem, t)s coming from object O, could then interact with appropriate machinery of the SMM(brain) of the sensitive, or, in some cases with the Shadow Matter eyes of the sensitive. In this way, the information derived from objects or people who had been close to an object O, for a sufficiently long time, could, via object O, i.e. via the SMM(complem, t)s emitted by the SMM(object) of an object O, be made available directly, or indirectly, to the SMM(brain) of suitable sensitives.

There has been much debate about the validity of some of the available putative evidence for object reading (see West, 1954, pp. 76-7) and some early experimental work by Hettinger was discredited on methodological grounds. Undoubtedly, many of Osty's findings and those of others relating to object reading may be questioned by hyper-sceptics. Yet, from the present theoretical point of view, the feasibility of object reading seems as plausible as the occurrence of apparitions for which the evidence is, perhaps, less ambiguous, notably in the case of haunted houses.

What is significant is that I have established a coherent theory, which, up to this stage already, explains out of the body experiences (OBEs), telepathy, clairvoyance, haunted houses and object reading within a single unified theoretical framework, which is entirely physicalistic and mechanistic.

The fact that the 'Committee for Scientific Investigation of Claims of the Paranormal' (CSICOP, see p. 64) 'wishes that people would not believe in ghosts or telepathy' is irrelevant.

2.6 Osty's discoveries of the properties of object reading

As Osty's works are not readily available, I shall cite from Tyrrell (1948 p. 176). He writes:

Osty, referring to one of his best sensitives, says: 'The many diverse experiments I have made with Mme Morel have taught me that the object placed in her hands avails to set her faculty in action, not by the fact of having belonged to such and such a person, but by having been touched by that person.' This was seen because, if an object belonging to A was brought by B and given to C for transmission to Mme. Morel, she would begin by describing C who had touched it last; then, if told to go back, would describe B and finally the owner of the object A. This makes the role of the object very puzzling. If touching the object is effective, one might think that the act of touching it impressed something on the material of which the object is made; but the summary of his observations, which Osty gives, would seem to discount this view. The most important of his observations are:

(1) After the sensitive, by holding the object, has once achieved psychological connection or rapport with the owner or contactor the object may be destroyed without affecting the sensitive's power to give information about him. Psychological rapport

with the contactor having once been established, information is sometimes given about events which happened to him after the object has been destroyed.

I believe that my theory can clear up this and other puzzling phenomena about object reading. Suppose A contacts object O. This could lead to SMM(complem, t[A]) being transferred to the SMM(object) of O. When the sensitive subsequently touches object O, then SMM(complem, t[A]) (or its replicated descendant) can transfer to, and interact with, the sensitive and his or her SMM(brain). The sensitive could then, via SMM(complem, t[A]) establish telepathic communication with person A. This could continue even after the object O is destroyed. It could convey information about events that happened after the destruction of object O.

(2) The material of which the object is made does not matter. Since all relevant interactions in this case are between Shadow Matter, the material of the object may not be relevant.

(3) If objects used in this way are allowed to touch one another, it does not make any difference.

This could simply mean that SMM(complem, t)s that are firmly attached to SMM(object) of one object cannot easily be transferred to SMM(object) of another object.

(4) The length of time during which the owner has possessed the object or made contact with it does not matter.

A possible explanation is that, although with increasing time of possession more copies of the same SMM(complem, t) could become 'fixed' to the same object, this does not matter, since each SMM(complem, t) that is fixed can replicate.

(5) The lapse of time since the owner last touched the object does not matter.

This could mean that SMM(complem, t)s that become captured by the SMM(object) of an object adhere to the object indefinitely, and have a long 'life-time'.

(6) When once the sensitive has entered into the life of the owner of the object, the whole of that life is accessible and not merely the portion of it during which he possessed the object.

When the owner of object O deposits an SMM(complem, t) on the SMM(object) of the object, then the latter, being complementary to the owner's SMM(body) (including its SMM(brain)), would contain representations of all memory traces of the owner, including memory traces formed before the owner acquired the object. This should enable the sensitive, who contacts a copy of SMM(complem, t), together with telepathy (see (1), p. 94) to have access to all of the owner's memories, and other cognitive constructs.

Notes to Chapter 2

1. Note that blood cells within an organ are not organ-specific cells.

2. As described, e.g. by Köhler and Wallach (1944) and criticized by Wassermann (1978) among others.

3. This footnote is partly for readers who are familiar with molecular genetics. Just as DNA can serve as a template for forming messenger RNA, whereas messenger RNA cannot form

DNA (in the absence of a so-called 'reverse transcriptase' (i.e. a specific viral enzyme)), so, I assume, that a SMM(complem) (which is analogous to a messenger RNA) cannot generate a SMM(object) (which is here analogous to a DNA). More generally, my theory assumes that only ordinary matter objects can act as templates for formation of SMM(object)s with just one SMM(object) per ordinary matter object. This ensures, for instance, that there is only one SMM(brain) in our region of the universe for any particular ordinary matter brain.

4. Basically my further assumption was suggested by an analogy with gene amplification in molecular biology. A single gene-representing piece of chromosomal DNA in a cell could not achieve much by itself, without 'gene amplification,' by permitting to be multiply represented by many similar messenger RNA copies made by that DNA in succession. Even this would not be enough. A further substantial amplification effect is achieved when each messenger RNA macromolecule, made by the gene, can, in turn, generate (with appropriate cellular machinery) a large number of copies of the same kind of polypeptide chain of a particular type of protein. In this way a single gene could generate, indirectly, a vast number of similar polypeptide chains.

5. The mechanism that I am postulating is analogous to that of DNA replication, where each strand of a two-stranded DNA double-helix can serve as a template for making the other, complementary, strand, in the presence of an appropriate polymerase. A polymerase is a protein molecule (enzyme) that helps in the step by step synthesis of a polymer for which it is a specific. (A polymer is an extended, string-like molecule that consists of a chain of strung together sub-units much like beads on a necklace.) In the Shadow Matter world there could also exist systems analogous to polymerases.

6. The above mechanism of clairvoyance is best suited to cases of travelling clairvoyance but is also applicable to all other cases of clairvoyance; see chapter 3.

7. It seems likely that all kinds of other objects, say houses, grass stalks etc., both inanimate and animate, could form and emit SMM(complem, object)s. These, however, might not play any part in the clairvoyant mechanism, but could represent SMM(object)-Junk.

In the study of gene-carrying systems, the so-called 'genomes' in plants and animals, scientists have found something analogous. A high percentage of a genome consists of 'genetic junk,' whereas only a small percentage of the genome represents 'genes'.

8. There are probably millions of people in the world who can play the piano. But only a very few can or could play it as well as Sviatoslav Richter. If we asked anyone who plays the piano to appear on TV, we would probably be as disappointed as Mr. Randi was with some of his mediums. But this does not mean that there are not people as gifted as Richter.

9. If, as is assumed here, the apparition is synthesized by the percipient's SMM(brain), but is induced by a Shadow Matter precursor, namely a SMM(complem, b), then this suggests the following. Apparently Shadow Matter precursors of apparitions can be passed through by ordinary matter as if that Shadow Matter does not exist. Much the same happened in an out of the body experience (Case 5). Describing **Case 5**, the percipient, Dr Wiltse, wrote (as cited on p. 47) that during his OBE 'As I turned, my left elbow (i.e. the left elbow of his SMM(body) which, according to the present theory had left Wiltse's ordinary matter body) came into contact with the arm of one of the two gentlemen who were standing in the door. To my surprise, his arm passed through mine without apparent resistance, the severed parts closing again without pain, as air reunites.'

10. See my remarks on psi-phenomena and hypnosis on p. 43.

11. I do not share Tyrrell's opinion that the letter could not have contained all the information given by the sensitive, although Osty would have noticed and reported that the sensitive read the letter.

3

Mechanistic Explanations of Some Cases of Clairvoyance and Telepathy

3.1 Interpretations of cases of clairvoyance

3.1.1 The indirect relationship between 'travelling clairvoyance' and 'out of the body experiences' : A Case Study

In section 2.2.1 (p. 80) I noted that the postulated mechanism of SMM(complem) amplification permits psi-communication over long distances. Moreover, just as radio transmission enables photons to be emitted from the transmitting antenna to a vast number of spatial regions, so SMM(complem)s, emitted by a psi-source, could, with the aid of SMM(complem) amplification, permit appropriate copies of themselves to reach almost any locality on Earth. I shall illustrate this, and additional aspects of this mechanism, by explaining a case of so-called 'travelling clairvoyance' in terms of the mechanism of section 2.4.

Case 15

This case, which involves a reciprocal apparition, was cited by Tyrrell (1953, pp. 116-7) and Myers (1903), and was originally reported in *The Proceedings of the Society for Psychical Research* (London) vol.7, pp. 41 ff.

Mr S. R. Wilmot, an American, was crossing the Atlantic in 1863, returning home in company with a friend, a Mr W. J. Tait who shared his cabin. The cabin was right aft, and owing to the slope of the ship's side the two berths were not vertically over one another, Mr Wilmot occupied the lower berth and Mr Tait the upper.

After eight days of bad weather, Mr Wilmot was enjoying his first night of refreshing sleep, when,' as he says, 'towards morning I dreamt that I saw my wife, whom I had left in the United States, come to the door of my stateroom clad in her night-dress. At the door she seemed to discover that I was not the only occupant of the room, hesitated a little, then advanced to my side, stooped down and kissed me and after gently caressing me for a few moments, quietly withdrew. Upon waking up I was surprised to see my fellow passenger leaning upon his elbow and looking fixedly at me. 'You're a pretty fellow,' said he at length, 'to have a lady come and visit you in this way.' I pressed him for an explanation. . . at length he related what he had seen while wide-awake, lying in his berth. It exactly corresponded with my dream. . .

The narrator says that, on arriving home and meeting his wife, almost her first question when we were alone together was, 'Did you receive a visit from me a week ago Tuesday?' 'A visit from you?' said I 'We were more than a thousand miles at sea.' 'I know it,' she replied, 'but it seemed to me that I visited you.' 'It would be impossible,' said I, 'Tell me what makes you think so.' 'My wife then told me that on account of the severity of the weather and the reported loss of the *Africa*. . . she had been extremely anxious about me. On the night previous, the same night when. . . the storm had just begun to abate, she had lain awake for a long time thinking of me, and about four o'clock in the morning it seemed to her that she went out to seek me. Crossing the wide stormy sea, she came at length to a low, black steamship, whose side she went up and then descended into the cabin, passed through it to the stern until she came to my stateroom. 'Tell me' she said, 'do they ever have staterooms like the one I saw where the upper berth extends further back than the under one? A man was in the upper berth looking right at me, and for a moment I was afraid to go in, but soon I went up to the side of your berth, bent down and kissed you and embraced you and then went away.'

Tyrrell added that 'Mr Wilmot's wife and sister add their testimony in confirmation of the report.' (At the time the report was communicated, Mr Tait had died, so that his testimony was not available; see Myers, 1903, pp. 682 ff.)

At first thought it might be tempting to explain this, extremely complex, 'reciprocal apparition' and other cases which seem to involve 'travelling clairvoyance' in which a percipient seems to travel to a 'target' (see Tyrrell, 1953, pp. 119-124) on the hypothesis that the mechanism involved is akin to that of an OBE. This would mean that, as in an OBE, Mrs Wilmot's SMM (body) (including her SMM(brain)) would have become detached from her ordinary matter body and moved across the sea towards the ship, ascended the ship's side and made its way to the stateroom and there had various cognitive experiences, such as 'being afraid of entering the stateroom,' and then being motivated to kiss and embrace Mr Wilmot. Such an explanation, however, seems implausible. According to my theory of OBEs, the detached SMM(body) after leaving the ordinary matter body of the percipient remains bound to the ordinary matter body by an elastic cord of Shadow Matter. It seems highly unlikely that such a cord would stretch over a distance of more than a thousand miles from Mrs Wilmot's ordinary matter body to the ship at sea. Worse, how could a single SMM(body) be directed accurately to the right target, i.e. the ship and the stateroom inside it, and 'know' that Mr Wilmot was to be found there? Moreover, unlike in most, or all, typical OBEs, Mrs Wilmot did not observe her own ordinary matter body from the outside after the start of her journey. Such considerations suggest that an OBE mechanism was unlikely to be instrumental in this case, and similar ones, of travelling clairvoyance.

Alternatively it could be assumed that Mrs Wilmot's SMM(body) remained attached to her ordinary matter body, but that the SMM(body) gave rise, sequentially, to a large number of SMM(complem)s. Each of these SMM(complem)s will be referred to as SMM(complem, Mrs W). According to my theory of clairvoyance of section 2.4, these SMM(complem, Mrs W)s would move off in random directions from Mrs Wilmot's ordinary matter body (or, rather, from its attached SMM(body)), and while moving, would replicate, leading to a SMM(complem, Mrs W) copy number amplification (see section 2.2.1, p. 80). Thus, what would have moved away from Mrs Wilmot's ordinary matter body was not her SMM(body) but complementary copies of that SMM(body), namely SMM(complem, Mrs W)s. Whenever a SMM(complem, Mrs W) replicates during motion, the resulting two SMM(complem, Mrs W) copies (one of which is the original) would tend to move off randomly along branching paths. Accordingly, one or more SMM(complem, Mrs W) copies could ultimately have reached the location of Mr Wilmot on board ship, via the pathway described later by Mrs Wilmot. After all, if radio communication had been used, photons from the location of a sender near Mrs Wilmot's body could also have reached Mr. Wilmot at sea, if the transmitter had been sufficiently energetic. But the proposed mechanism is, of course, totally different from transmission via photons. Yet, in order to see, to embrace and kiss Mr Wilmot, to see Mr Tait in the stateroom, etc., it may be assumed, according to the present theory, that the SMM(brain)s of all SMM(complem, Mrs W) copies (called simply SMM(complem, brain, Mrs W)) are capable of cognitive transactions to the same extent as SMM(brain) of Mrs Wilmot's SMM(body) (which remained attached to her ordinary matter body).[1] Moreover, what each SMM(complem, Mrs W) had registered cognitively (but not necessarily consciously) during its movement is assumed to be passed on to its direct or indirect descendant SMM(complem, Mrs W) copies in the SMM(complem) replication process in the form of memory traces.

Ultimately some of the descendants of that SMM(complem, Mrs W) which was involved in the process of seeing (via its Shadow Matter eyes) and feeling the embracing and kissing of Mr Wilmot's SMM(body) (by means of the SMM(body) of the SMM(complem, Mrs W)), or that SMM(complem, Mrs W) itself, could via random movements, combined with repeated replications, reach Mrs Wilmot's SMM(body). An interaction between an appropriate copy of SMM(complem, Mrs W) and Mrs Wilmot's SMM(body) could then occur. During this interaction this appropriate copy of SMM(complem, Mrs W) could lead to the formation of memory traces by Mrs Wilmot's SMM(brain) of memory traces present on the interacting copy of SMM(complem, Mrs W) which was originally derived from the interaction of an appropriate copy of SMM(complem, Mrs W) with the SMM(body) of Mr Wilmot etc., described above. Likewise, memory traces formed by SMM(complem, Mrs W) during its journey to the ship or of the movement along the ship, or of its movement to the stateroom, etc., could induce corresponding memory traces in Mrs Wilmot's SMM(brain), by the mechanisms already postulated and described. As a consequence of this Mrs Wilmot's SMM(brain) could then enact the memory trace sequence of all the cognitive events reported in her case history cited above. It remains to explain how Mr Wilmot and Mr Tait recognized the embracing and kissing apparitions of Mrs Wilmot in the stateroom. Perhaps the simplest explanation is to assume that a copy of SMM(complem, Mrs W), derived via one or more replications from a primary copy, interacted with SMM(body) of Mr Wilmot. This could have triggered off a phantasy in both the SMM(brain) of Mr Wilmot and the SMM(brain) of SMM(complem, Mrs W), both phantasies being of the same kind. That such phantasy-generating mechanisms, leading to a very intricate apparitions, exist can be seen from **Case 3a**, where such a phantasy was hypnotically induced (p. 42) and appeared post-hypnotically. It should be stressed that, as far as Mr Wilmot and Mr Tait were concerned the apparition was a collective apparition. As in Case 1 it could be a telepathic apparition. The phantasy elaborated by Mr Wilmot's SMM(brain) could by the telepathic mechanisms already suggested in section 2.3 (p. 85) have induced a corresponding apparition in Mr Tait's SMM(brain). One also has to assume that the SMM(brain)s of Mr Wilmot and Mr Tait were in receptive states, which, according to my theory involves partly weakened bonding between the ordinary matter brain and the SMM(brain).

Another interesting example of travelling clairvoyance (where a percipient, as in the case of Mrs Wilmot above, apparently travels to another locality and reports what he or she found there) can be found in Mrs Sidgwick's paper *Proceedings of the Society for Psychical Research* (London) vol.7, pp. 58 ff). The case was abbreviated by Tyrrell as follows, and shows the class characteristics of travelling clairvoyance which are partly also present in the Wilmot case (**Case 15**).

Case 16

Tyrrell wrote:
Such a subject was 'Jane' the wife of a Durham pitman, for whose case Myers collected evidence. 'She never received any fee,' he writes, 'or made any exhibition of her powers.' Jane appears to have fallen spontaneously into a peculiar kind of trance in which she was able to answer questions put to her. On one occasion Dr F., the operator, told a patient of his, a Mr Eglinton, that he would try the experiment of getting Jane to visit him clairvoyantly between 8 and 10pm and Mr Eglinton said he would be in a particular room at the time. He was very thin from the effect of an illness, and at the appointed time Jane, in trance, was led in thought to the house by verbal direction and guided to the right room. She then said that she saw the door opening, and a very fat man with a corporation[2] and a cork leg coming in, and sitting down at the table with papers beside him and a glass of brandy and water at his side. Asked if she could see his name on any letters about, she said Yes; she spelt it correctly - Eglinton. Dr F thought the description of the man completely wrong, but found afterwards that Mr Eglinton had wished to try and experiment and had his clothes stuffed with pillows to represent a very fat figure and placed at the table, on which were papers and a glass of brandy and water.

Jane's trance could, according to my theory, be the result of weakened bonding between her ordinary matter brain and her SMM(brain), and this could have facilitated psi-activity of her SMM(brain) (see p. 38-9). The mechanisms involved in the travelling clairvoyance of this case could, with a few alterations, be of the same kinds as those involved in **Case 15**. There was also a measure of creative thinking involved, since Jane saw the fat man 'coming in,' which was not the case.

3.1.2. Travelling clairvoyance combined with an out of the body experience

The following case, taken from Hart's (1956, p. 175) paper, concerns an OBE which is combined with travelling clairvoyance.

Case 17

Walter E. McBride was a bachelor farmer whose address was Indian Springs, Indiana, Route 1. He stated that on 23 December 1935 he had been wondering about his father during the entire day, and was under the impression that he might be ill. Shortly after retiring about eight o'clock that evening, he reportedly found himself floating in the room, in a whitish light which cast no shadow.[3] He says he was wide awake at the time. He reportedly found that he was floating upwards through the building; the ceiling and floor failed to stop him. After reaching a certain height his body turned vertical, and looking downwards he saw his physical body upon the bed. [The reader should note the close similarity of some of the class characteristics of this OBE and that of **Case 7a**, reported by Muldoon.]

Almost at once he realized that he was moving through the air towards the north and he seemed to know that he was going to his old home several miles away. Passing through the walls of his father's house he stood at the foot of the bed in which he saw his father reclining. His father's eyes were fixed upon him and he seemed surprised, but did not seem to hear when McBride spoke to him. The knowledge came to him that his father was well, whereupon he found himself travelling back to his bedroom, where again he saw his own body, still lying on the bed where he had left it. Upon re-entering his physical self he was instantly alert, with no feeling of drowsiness. Throughout his projected excursion, McBride was aware of a presence, which he was unable to identify, but which he subsequently came to regard as a guide.

Immediately upon recovering possession of his physical body, McBride got up, made a light, and wrote down an account of what he had just experienced. Two days later, on Christmas Day, he visited his father, who verified his experience, by saying he had seen McBride, just as he stood at the foot of the bed. The father, moreover, had written down the time of his vision, and it tallied with the time written down previously by the excursionist.

The early phase of this case could be interpreted as a straight forward OBE, and, likewise, the last phase. Typically in these phases the percipient McBride (Jr.) saw his ordinary matter body upon the bed from a position outside that body (and also saw the white light that, in the early stages of an OBE, also characterizes some other OBEs, described by Moody (1976), see p. 37). According to my theory, detachment of McBride's SMM(body) (including his SMM(brain)) during his OBE would facilitate psi-phenomena (see p. 38-9). The upward floating experience of McBride during part of the OBE phase may be compared with, and is similar to, that of Bertrand's upward floating (see **Case 7**, p. 54 ff ; see also **Case 4**, p. 44, and **Case 7a** p. 65). Again, like Bertrand, McBride combined apparent clairvoyance with an OBE (e.g, when Bertrand recognized his wife with a party of others at the hotel at Lugern (see p. 55) which was totally out of his sight). Moreover, McBride's 'visit' to his father and his recognition of the apparition of his father, at that time, seems to involve travelling clairvoyance with reciprocal recognition of a kind strikingly similar to that of the Wilmot case (**Case 15**). Accordingly, for the travelling clairvoyance part of the reciprocal recognition of the McBride case, the theory of **Case 15** could apply *mutatis mutandis*. The similarity of the present case and of Case 15, as regards the travelling clairvoyance aspects also exemplifies that class characteristics of quite complex psi-phenomena occur repeatedly and spontaneously. My comparisons of Case 7 and the present case also lead to similar conclusions.

3.1.3 Mechanistic interpretations of miscellaneous cases of clairvoyance, object reading and hypnosis

One of the best early case collections of clairvoyance, and perhaps still the best collection in existence, is due to Mrs Henry Sidgwick, wife of one of the founders of the Society for Psychical Research (London) (Sidgwick, 1891). Many of the case histories are accompanied by multiple testimonials from people who were often (but not always) told about the reported events near the time of their occurrence. Yet, even if we make allowance for occasional possible slight misrepresentations of minor details of the evidence, then there remains the fact that several of the case histories have their parallels in other case histories reported in the two volume classic *Phantasms of the Living* by Gurney *et al.*(1886). If die-hard sceptics try to shrug off such case histories as due to faulty memories, misrepresentations on a grandiose scale and similar clichéd arguments, simply because they are convinced that such phenomena could not happen, owing to the alleged inconceivability of explanatory physical mechanisms, then I suggest that such people are self-deluding, as self-deluding as the many who did not accept the existence of hypnotic phenomena in the nineteenth century (see also my remarks in section 1.11). Let me start with the experience of Mrs Agnes Paquet, reported in 1890; for the full case history see Sidgwick (1891), pp. 32 ff.

Case 18

At the beginning there occurred the following accident.

On October 24th 1889, Edmund Dunn, brother of Mrs Agnes Paquet, was serving as fireman of the tug *Wolf*, a small steamer engaged in towing vessels in Chicago Harbour. At about 3am the tug fastened to a vessel, inside the piers, to tow her up the river. While adjusting the tow-line Mr Dunn fell or was thrown overboard by the tow line, and drowned. The body, though sought for, was not found until about three weeks after the accident, when it came to the surface near the place where Mr Dunn disappeared.

On the morning of the day of the accident Mrs Agnes Paquet awoke gloomy and depressed. Later in the morning she went into the pantry and stated that 'as I turned round my brother Edmund - or his exact image - stood before me and only a few feet away. The apparition stood with back toward me - seemingly impelled by two ropes drawing against his legs. The vision lasted but a moment, disappearing over a low rail or bulwark but was very distinct. I dropped the tea, clasped my hands to my face, and exclaimed, "My God Ed is drowned!"' ... 'At about half past ten am my husband received a telegram from Chicago announcing the drowning of my brother.' Mr Paquet thought that Edmund was sick in hospital in Chicago, but Agnes Paquet replied 'Ed is drowned; I saw him go overboard,' and she gave her husband a careful description of the apparition she had seen and described the appearance of the boat at the point where her brother went overboard. She had noticed that her brother was bareheaded, had on a heavy blue sailor's shirt, no coat and that he went over the rail or bulwark. She also noticed that his pants' legs were rolled up enough to show the white lining inside. All this was confirmed by the crew afterwards, except that they thought that Mr Dunn had his hat on at the time of the accident. The crew said that Mr Dunn had purchased a pair of pants a few days before the accident occurred, and as they were a trifle long before, wrinkling at the knees, he had them rolled up, showing the white lining as seen by Mrs Paquet.

Mrs Sidgwick (1891) stressed that Mrs Paquet's experienced apparition did not coincide with the time of death, but occurred six hours afterwards. This case could be explained again in terms of the mechanisms of clairvoyance of p. 87, which were already invoked in **Cases 15-17**. It may be assumed that Mrs Paquet's SMM(body) emitted many SMM(complem)s which moved off randomly and replicated while travelling, thus enabling the replicas to reach most or all loci on Earth. One of the replicas of SMM(complem) could then have reached the neighbourhood of Mr Dunn and the part of the ship on which he was located, at the time of the accident. Sphotons (i.e. Shadow Matter photons) emitted by the SMMs of Dunn's face, suit, etc. and by the ship near him could then have acted on the SMM(complem) leading it to form memory traces of the accident. Subsequently that SMM(complem) moved on and other replicas were formed, one of which reached Mrs Paquet's SMM(body), where the SMM(brain) of SMM(complem) could have interacted with the SMM(brain) of Mrs Paquet's SMM(body). This could have led to transactions of Mrs Paquet's SMM(brain) which represented the accident, without, however, conscious awareness of this. This representation could have generated a memory trace sequence of the accident. This unconscious representation could have led to a feeling of depression and gloom which Mrs Paquet experienced on awaking from her dream. The full effect of the 'captured' SMM(complem) may then only have surfaced many hours later, when the memory trace sequence formed was reactivated. There is nothing particularly far-fetched about this explanation, since even in ordinary perception many events can register unconsciously long before they surface fully into awareness (possibly also by reactivation of memory traces formed earlier of the perceived events or things). Such delayed action happened also, typically, in the long-delayed post-hypnotic apparition described by Gindes (see Case 3a, p. 42), where the nature and time of appearance of the apparition was suggested hypnotically many months before it occurred.

A case similar to the Paquet case, involved an apparition appearing in a dream, which was preceded by a feeling of nervousness all evening, and which related to the death of a brother. The case is described by Mrs Storie in *Phantasms of the Living* (Gurney *et al* 1886) vol.1, p. 370 No. 134). Yet, as Mrs Sidgwick notes, in this case the nervousness began before the accident occurred, and could have been related to precognition of the events. Mrs Sidgwick referred also to case no. 65 of Gurney *et al.* (1886, vol.1, p. 268) which exhibited also depression and deferment of the surfacing of the apparition, So, we are dealing here, once more, with repeatable class characteristics of psychic phenomena.

Let me turn to another case of clairvoyance, recorded by Mrs Sidgwick (1891).

Case 19

In Washington DC, January 14th 1889 between 2 and 3pm Mrs Conner is going up the steps of her residence No. 217, Delaware-Avenue, carrying some papers. She stumbles, falls is not hurt, picks herself up, enters the house.

At about the same time - certainly within the hour, probably within 30 minutes, perhaps at the very moment - another lady. . . Mrs B, is sitting sewing in her room about 1½ miles distant. The two ladies are friends but had not met this day. Mrs B 'sees' the little accident in every detail. The vision or image is minutely accurate (as it afterwards proves). . .

The apparition was so vivid that Mrs B sat down at once and wrote a letter to Mrs Conner. She wrote:

'I was sitting in my room sewing, this afternoon about two o'clock, when what should I see but your own dear self, but heavens in what a position. . . You were falling up the front of the steps in the yard. You had on your black skirt and velvet waist, your little straw bonnet, and in your hand were some papers. When you fell your hat went in one direction and the papers in another. You got up very quickly, put on your bonnet, picked up the papers, and lost no time getting into the house. You did not appear to be hurt, but looked somewhat mortified.'

All this turned out to be correct. This case raises an aspect of some, perhaps most or all, apparitions of the living. Unlike some other apparitions which were only perceived statically for a few moments, this apparitions, in common with many others (see **Case 1**, p. 24), seemed to show complex dynamic behaviour, including sequences of movements of a person or of people. This, however, does not create any serious problems for the present theory, since SMM(complem)s could perceive moving objects, and form memory traces of these movements, in much the same way as SMM(brain)s of people. Also, in the present case Mrs B's SMM(body) could have synthesized SMM(complem)s and the remainder of the explanation could be, with suitable adjustments, as in **Case 18**.

Case 20

Another case of apparent clairvoyance from Mrs Sidgwick's 1891 collection of such cases was reported by Mr H. M. Lee, son of the late Dr Henry Lee,

Bishop of Iowa, in a letter to a relative, dated Syracuse, NY December 16th 1887. Apparently Mr Lee was often 'psychically' aware when his father was in danger, even when they were separated by many miles. For instance, on a certain night in 1874 the Bishop fell downstairs. The same night Mr H. M. Lee, who was very tired, went to bed, fell asleep and did not even hear his wife come to bed. He writes that in his sleep he 'knew nothing till I saw father at the top of the stairs in the act of falling. I jumped to catch him and landed on the floor on my feet, with considerable noise. My wife awoke and wanted to know what on earth I was trying to do. I by that time had lighted a lamp, and upon looking at my watch found it was quarter-past two. I asked my wife if she heard the crash. She said no. I then told her what I had seen, but she tried to laugh me out of it, not succeeding, however. . . Early in the morning I went to town and telegraphed home, inquiring if all was well, and received a letter from father which fully corresponded with my visitation to the very minute. . . '

In fact, the Bishop had for the first time occupied a new residence built for him by the diocese.

Not being accustomed to the interior arrangements, he one night took a false step, turning towards the stairway instead of his own room, and fell down the stairs, a flight. . . of twenty steps.

This is cited from a letter by E. Sullivan of December 29th 1887 to the Society for Psychical Research, and appears in Mrs Sidgwick's paper. Here, again, we have a clairvoyantly perceived movement sequence as in **Case 19**. A similar case is cited by Gurney *et al.* (1886, vol.1, p. 338 No. 108), showing that such cases have repeatable class characteristics. The explanation of the present case, in terms of SMM theory, could, therefore, be similar to that given for Case 19, and need not be repeated (with modified details).

Case 21

Another startling case, found in Mrs Sidgwick's (1891) paper, was contributed by Dr Golinski, who practiced medicine at Krementchug in Russia. One day in July 1888 he had his usual after luncheon sleep and dreamt, as he writes:

that the door bell rang, and that I had the usual rather disagreeable sensation that I must get up and go to some sick person. Then I found myself transported directly into a little room with dark hangings. To the right of the door leading into the room is a chest of drawers, and on this I see a little paraffin lamp of a special pattern. I am keenly interested in the shape of this little lamp, different from any it has previously happened to me to see. To the left of the door I see a bed, on which lies a woman suffering from a severe haemorrhage. I do not know how I come to know that she has a haemorrhage, but I know it. I examine her, but rather to satisfy my conscience than for any other reason, as I know beforehand how things are, although no one speaks to me. . .

About ten minutes after I awoke the door bell rang, and I was summoned to a patient. Entering the bedroom I was astonished, for I recognized the room of which, I had just dreamt. The patient was a sick woman, and what struck me especially was the paraffin lamp placed on the chest of drawers exactly in the same place as in my dream, and of the same pattern, which I had never seen before. My astonishment was so great that I, so to speak, lost the clear distinction between the past dream and the present reality, and, approaching the sick woman's bed, said affirmatively, 'You have a haemorrhage,' only recovering myself when the patient replied, 'Yes, but how do you know it?'

Dr Golinski awoke from his dream at 4.30 and the patient had decided at 4.30 to send for him.

Here, again Dr Golinski's SMM(body) could have generated SMM(complem)s. These, via spatial (random) propagation and replication, could have generated descendant SMM(complem)s some of which reached the patient and her surroundings and, as was suggested in other cases, could via sphotons emitted by the patient's SMM(body) and emitted by SMM(object)s of objects in her surroundings, act on the SMM(complem)s. One of these or one of its descendants could then have reached Dr Golinski's SMM(body) and produced the clairvoyant recognitions by interacting with the SMM(body). Since my theory assumes that SMM(brain)s operate similarly in normal and paranormal perception, we could expect that the SMM(brain) can use the same mechanisms of attention and 'cue selection' or discrimination or 'psychological set' in psi-perception as in normal perception.

Case 22

[and its relevance to *a theory of dream states*]

Next let me cite and discuss another case from Mrs Sidgwick's (1891) collection. It was reported by Margaret R. Wedgwood in February 1884. She had a remarkably vivid dream which she told next morning to her father-in-law. She writes:

I dreamt I went to a strange house standing at the corner of a street. When I reached the top of the stairs I noticed a window opposite with a little coloured glass, short muslin blinds running on a brass rod. The top of the ceiling had a window veiled by gathered muslin. There were two small shrubs on a little table. The drawing-room had a bow window, with the same blinds; the library had a polished floor, with the same blinds.

As I was going to a child's party at a cousin's, whose house I had never seen, I told my father-in-law I thought that that would prove to be the house.

On January 10th I went with my little boy to the party, and, by mistake, gave the driver the wrong number. When he stopped at No. 20, I had misgivings about the house, and remarked to the cabman that it was not a corner house. The servant could not tell me where Mrs. H. lived, and had not a blue-book. Then I thought of my dream, and as a last resource I walked down the street looking up for the peculiar blinds I had observed in my dream. These I met with at No. 50, a corner house, and knocking at the door, was relieved to find that it was the house of which I was in search.

On going upstairs the room and windows corresponded exactly with what I had seen in my dream, and the same little shrubs in their pots were standing on the landing. The window in which I had seen the coloured glass was hidden by the blind being drawn down, but I learned on enquiry that it was really there.

It is interesting that the present **Case 22**, like **Cases 20** and **21**, apparently involved clairvoyant awareness during dream states of the percipient. This, according to my theory, suggests that dream states may help to facilitate the occurrence of psi-phenomena. I postulated already (p. 39) that states of weakened binding of SMM(brain)s to ordinary matter brains could facilitate the responses of SMM(brain)s to telepathy - and/or clairvoyance-mediating signal systems (e.g. SMM(complem)s). This suggests that in dreams, as in mediumistic trance, there also occurs a weakening of the bonding of SMM(brain) and ordinary matter brain. Yet, whereas I assumed that in mediumistic trance there occurs complete breakage of many Shadow Matter-ordinary matter bonds between SMM(brain) and ordinary matter brain (see p. 40), in dreams there may occur no breakage only weakening of bonds, due to changes of states of constituents of the ordinary matter brain. This weakening of bonds could ensure a greater autonomy of the SMM(brain), leading to facilitation of SMM(complem) access to, and interaction with, the SMM(brain). Nevertheless, dreams per se are unlikely to be sufficient to ensure the occurrence of psi-phenomena, as judged by the vast number of dreams that occur every night for more than a billion people, most of whom do not seem (or claim) to have any psychic experiences during their dreams. It could be argued that many of these psychic experiences are not reported. This, however, remains unsubstantiated.

Case 23

Let me cite now a case described by the great philosopher Immanuel Kant (see 1900, pp. 157-8), concerning an experience by Swedenborg:

In the year 1759 towards the end of September, on Saturday at four o'clock pm, Swedenborg arrived at Gothenburg from England, when Mr. William Castle invited him to his house, together with a party of fifty persons. About six o'clock Swedenborg went out, and returned to the company quite pale and alarmed. He said that a dangerous fire had just broken out in Stockholm, at the Södermalm, and that it was spreading very fast. He was restless and went out often. He said that the house of one of his friends, whom he named, was already in ashes, and that his own was in danger. At eight o'clock, after he has been out again, he joyfully exclaimed, 'Thank God! the fire is extinguished; the third door from my house.' This news occasioned great commotion throughout the whole city, but particularly among the company in which he was. On Sunday morning Swedenborg was summoned to the governor who questioned him concerning the disaster. Swedenborg described the fire precisely, how it had begun and in what manner it had ceased, and how long it had continued. On the same day the news spread through the city, and as the governor thought it worthy of attention the consternation was considerably increased; because many were in trouble on account of their friend's property which might have been involved in the disaster. On Monday evening a messenger arrived at Gothenburg, who was despatched by the Board of Trade during the time of the fire. In the letters brought by him, the fire was described precisely in the manner stated by Swedenborg. On Tuesday morning the royal courier arrived at the governor's with the melancholy intelligence of the fire, and the loss which it had occasioned, and of the houses it had damaged and ruined, not in the least differing from that which Swedenborg had given at the very time when it happened; for the fire was extinguished at eight o'clock.

What seems interesting in this case is that fire, involving lit gases, can be perceived clairvoyantly. In this case my theory suggests that Swedenborg's SMM(body), including his SMM(brain), emitted large numbers of SMM(complem)s many of which or their descendants reached Stockholm and interacted with the sphotons given off by the SMM(objects)s of the objects that were burning. The SMM(complem)s involved could then have formed memory traces of various parts of the fire. Many of these SMM(complem)s, or their descendants by replication could have returned to Gothenburg (by random motions) and some of them could have interacted with Swedenborg's SMM(body), including his SMM(brain).

Case 24

The following is the account of another case of 'travelling clairvoyance' reported by Dr Alfred Backman of Kalmar, Sweden (see Backman 1892). As summarized by Tyrell (1953, pp. 121 ff):

Backman experimented with a [hypnotized] subject rather similar to Jane [**Case 16** p. 100], called Alma. On one occasion without anything being prearranged, he told Alma to go to the Director-General of pilotage at Stockholm and see if he was at home. She described him as sitting at table in his study writing. Among other things she saw a bunch of keys on the table, and she was sharply ordered to seize and shake them and to put her hand on the Director-General's shoulder. This she repeated two or three times until Alma at last declared that he observed her. On being told of the experiment subsequently, the Director-General's account of his own experience was as follows. 'On that occasion,' he said, 'he was sitting, fully occupied with his work, when, without any reason whatever, his eyes fell on the bunch of keys, lying near him on the table. He then began to consider how he could have put the bunch of keys there, and why it was there, when he knew for certain that he was never in the habit of leaving it there. While reflecting on this, he caught a glimpse of a woman. Thinking it was his maid-servant, he attached no importance to it, but when the occurrence was repeated, he called her and got up to see what was the matter. But he found nobody, and was informed that neither his servant nor any other woman had been in the rooms. He did not observe any rattling of the bunch of keys or any other movement of the keys.'

This case shares several of the class characteristics of **Case 15**, and could be explained, with some adjustments, in terms of similar kinds of Shadow Matter mechanisms. An interesting difference between the present case and Case 15 is the use of hypnosis in the present case, whereas Case 15 occurred spontaneously, without any of the participating subjects being hypnotized. Since according to my earlier suggestions (p. 40) in hypnosis there could occur an abnormally tight binding between the ordinary matter brain and the SMM(brain), this would, at first sight seem to hyper-inhibit psi-signals (e.g. SMM(complem)s) from acting effectively on the SMM(brain). Yet, whereas hypnosis could lead to the setting up of particular states of the SMM(brain) via the ordinary matter brain, in the present case with the SMM(brain) related to the travelling clairvoyance, the abnormally tight coupling between the SMM(brain) and ordinary matter brain need not necessarily exclude responses of some parts of the SMM(brain) to incoming SMM(complem)s, since some parts of the SMM(brain) could be weakly coupled to the ordinary matter brain even in hypnosis. That hypnosis, as such, does not automatically assist in the occurrence of psi-phenomena seems to be born out by the fact that only rarely do psi-phenomena occur is hypnotized subjects. It could be that Alma's SMM(brain) was more weakly bonded to the ordinary matter brain during hypnosis than is the case with most ordinary hypnotized subjects.

Case 25

The following case, which also involves hypnosis combined with travelling clairvoyance, was reported by William Reid, a local correspondent of the *Aberdeen Journal*, who on 8th May 1850 cited the case, which was fully repeated in Mrs Sidgwick's (1891) paper (her p. 49), and shows that Cases 24 and 25, and other cases of 'travelling clairvoyance' share important class characteristics (i.e. that these class characteristics turn up repeatedly in different cases). Apparently on 23rd of April 1850 a lad was hypnotized and while in the hypnotic state was asked to be 'transported' (by travelling clairvoyance) to the icy regions and to report there on the ships *Hamilton Rose* and *Eclipse*. He stated in his answer:

That the first ship which would arrive here this season would be the *Hamilton Rose*, and that he at that moment saw the captain and surgeon of the vessel engaged in dressing the hand of the second mate, Cardno, who, he said, had accidentally lost part of some of his fingers. . .

On 3rd May the first whaler of the season arrived and proved to be the *Hamilton Rose*. The second mate Cardno was soon recognized with one of his arms in a sling. It turned out that he had accidentally shot away portions of some of his fingers when fishing.

As stressed, what is significant about the several cases of travelling clairvoyance, cited and explained in terms of my theory, is the repeatability of class characteristics. We have seen already that many out of the body experiences (OBEs) also show repeatability of their class characteristics (see also Moody, 1976). This endorses the points made in section 1.10 (notably on p. 59) that repeatability in parapsychology is a repeatability of class characteristics of phenomena and not, necessarily, a repeatability of the minutiae of details of observations. (The explanation of **Case 25** could in parts be similar to that of **Case 15**.)

Above all it must be obvious by now to even the most dogmatic of critics that the conditions in which genuine psychic phenomena happen in case after case cited (and yet to be cited below), differ drastically, in fact unrecognizably from the conditions demanded by James Randi (1991) in his 'tests' of psychic subjects. Genuine phenomena just do not take place at the demands of a magician, because in most cases they are spontaneous. Even **Cases 24, 25** (and **26**) below do not, in the least resemble the 'experiments' of Mr Randi. Like naturalists in biology, we are dealing here with psi-phenomena in the wild. It remains, of course, possible that sometimes psi-phenomena can be captured in the laboratory on a significant scale, although I prefer to study parapsychology in the wild. I should not be surprised if Mr Randi will deny the significance of every one of approximately eighty case histories cited in this book, because they do not suit, obviously, his preconceived doctrines. But I am not sure whether Mr Randi is used to scientific thinking, particularly thinking in the philosophy of science to which his book (Randi, 1991) does not belong.

Case 26

Another case involving the use of hypnosis combined with object reading (see sections 2.5 and 2.6) also comes from Mrs Sidgwick's (1891, p. 65) paper. It concerned a Mr A. W. Dobbie, a hypnotist, and two subjects who were hypnotized by him and, under hypnosis, gave clairvoyant accounts of a missing gold sleeve-link, whose fellow sleeve-link had been handed to Mr Dobbie by the Hon Dr Campbell. Dr Campbell told Mr Dobbie 'that he had no idea what had become of the missing sleeve-link, and asked Mr Dobbie to give the remaining gold sleeve-link to one of the clairvoyants.' One of these clairvoyants, Miss Eliza, who joined hands with her sister Miss Martha Dixon, the latter holding the sleeve-link, commenced as follows, viz:

I am in a house upstairs. I was in a bathroom, then I went into another room nearly opposite, there is a large mirror, just inside the door on the left hand, there is a double-size dressing table with drawers down each side of it, the sleeve-link is in the corner of the drawer nearest the door. When they found it they left it there. I know why they left it there, it was because they wanted to see if we could find it. I can see a nice easy chair there, it is an old one. . . it is nice and low. The bed has curtains, they are a sort of brownish net and have a fringe of dark brown. The wall paper is of light brown colour. There is a cane lounge there and a pretty Japanese screen behind it, the screen folds up. There is a portrait of an old gentleman over the mantelpiece, he is dead, I knew him when he was alive, his name is the same as the gentleman who acts as Governor when the governor is absent from the colony.[4] I will tell you his name directly it is the Rev Mr Way. It was a little boy who put the sleeve-link in that drawer, he is very fair, his hair is almost white, he is a pretty little boy, he has blue eyes and is about three years old. The link has been left on that table, the little boy was in the nursery, and he went into the bedroom after the gentleman had left. I can see who the gentleman is, it is Dr Campbell. . . Now I can hear someone calling upstairs, a lady calling two names, Colin is one and Neil is the other, the other boy is about five years old and is darker than the other. The eldest, Colin, is going downstairs now, he is going into what looks like a dining-room, the lady says, 'Where is Neil?' 'Upstairs ma,' 'Go and tell him to come down at once.' The little fair haired, boy had put the link down, but when he heard his brother coming up, he picked it up again, Colin says 'Neil you are to come down at once.' 'I won't' says Neil. 'You are a goose' replied Colin and he turned and went down without Neil. What a young monkey! Now he has gone into the nursery and put the link into a large toy elephant, he put it through a hole in front which is broken. He has gone downstairs now, I suppose he thinks it is safe there.

Now that gentleman has come into the room again and he wants that link; he is looking all about it, he thinks it might be knocked down: the lady is there now too, and they are both looking for it. The lady says - 'Are you sure you put it there?' the gentleman says, 'Yes'. Now it seems like next day, the servant is turning the carpet up and looking all about for it; but can't find it.

The gentleman is asking that young Turk if he has seen it, he knows that he is fond of pretty things. The little boy says, 'No'. He seems to think that it is fine fun to serve his father like that.

Now it seems to be another day and the little boy is in the nursery again, he has taken the link out of the elephant, now he has dropped it into that drawer, that is all we have to tell you about it, I told you the rest before.

Dr Campbell reported (see Sidgwick, 1891, pp. 66-7) that the conversation between the two children (and their names) are correct. The description of the room is accurate in every point. The portrait is that of the late Rev James Way. The description of the children is accurate. The link was discovered in the drawer between one sitting and the final one, and was left there pending the discovery of it by the clairvoyant. Neither of the clairvoyants had ever been to Dr Campbell's residence, and his children were utterly unknown to them, either in appearance or in name. Dr Campbell adds that the clairvoyants had no knowledge of his intention to place the link in their possession, or (more questionably) even of his presence at the seance, since both were hypnotized when he arrived.

Case 26, and many cases already cited, show that able sensitives or percipients in spontaneous cases do not guess wildly. The amount of detail given and its accuracy is totally at variance with the wild guessing assumption favoured by James Randi (1991). I believe that his wishful thinking exceeds his familiarity with case histories of psychic phenomena. By arguing that these case histories are only anecdotal some people believe that they can escape the tedious labour of careful study of highly relevant evidence. They use a cliché to get rid of what does not suit them. In this way they brush aside precise information given by first rate sensitives and use far less apt so-called sensitives to show that such accurate phenomena do not happen with them.

We have now seen in **Cases 24-6** that hypnotized subjects can, notwithstanding the assumed extra-tight linkage between ordinary matter brain and SMM(brain) of the hypnotized subjects, nevertheless be sensitive to psi-signals. In **Case 26** the sleeve-link handed to the clairvoyant subject may also have helped to bring about object reading, conceivably by the mechanisms suggested on pp. 93-4 of section 2.5. Case 26 could then be explained as follows. The SMM(brain) of the sensitive could, in the hypnotic state, emit SMM(complem)s of that SMM(brain) at an abnormally high rate. These SMM(complem)s will be referred to as SMM(complem, brain)s. Some of the emitted SMM(complem, brain)s could then impinge on the nearby sleeve-link (held by the sensitive) and interact with SMM(complem)s bound to the SMM(object) of the sleeve-link. Some of these sleeve-link bound SMM(complem)s could have been formed while the sleeve-link was originally in the vicinity of the other 'missing' sleeve-link of the pair. Hence, these captured SMM(complem)s could encode the events that concerned the 'missing' sleeve-link and the people associated with it. Also SMM(complem, brain)s could then absorb replicas of the SMM(complem)s associated with the sleeve-link and some of these SMM(complem, brain)s, or their replicas, could reach then again the SMM(brain) of the clairvoyant and interact with that SMM(brain) (which they match) and give off the absorbed SMM(complem)s (or replicated copies of the latter) that were derived from the sleeve-link. The SMM(brain) of the clairvoyant could then construct with the help of the received, sleeve-link-derived, SMM(complem)s, an account of what happened in the case reported. The cognitive machinery involved for constructing that account could be the same (according to my theory) as the machinery involved in normal cognitive construc-tions.

The assumption that the SMM(brain) may be hyperactive during hypnosis in emitting SMM(complem)s and may be more selectively 'set' is no more far-fetched than the assumption that during hypnosis the ordinary matter brain becomes specifically 'set,' as the following case shows.[5] ('Set' refers to being directed.)

Case 27

Mason (1952) who described a case of Ichthyosiform Erythrodermia of Brocq, and his treatment of it by hypnosis, notes that:

> The condition of Ichthyosiform Erythrodermia of Brocq is resistant to all forms of treatment [at least in 1952]. As a rule the course is progressive thickening of the skin from birth, reaching a maximum at the age of 15 and then remaining static throughout life. . .

Mason's case report is as follows:

> The patient was a boy aged 16 who suffered from congenital ichthyosis. The lesion consisted of a black horny layer covering the entire body except his chest, neck and face. The skin was papilliferous, each papilla projecting 2-6mm above the surface, and the papillae were separated from each other by only a very small distance, perhaps 1mm. . . The ichthyosiform layer, when cut, was of a consistent cartilage and was anaesthetic for a depth of several millimetres. . . On February 10, 1951, the patient was hypnotized and, under hypnosis, suggestion was made that the left arm would clear. (The suggestion was limited to the left arm so as to exclude the possibility of spontaneous resolution.) About five days later the horny layer softened, became friable, and fell off. The skin underneath was slightly erythematous, but normal in texture and colour. From a black and armour-like casing, the skin became pink and soft within a few days. Improvement occurred first in the flexures and areas of friction, and later on the rest of the arm. The erythema faded away in a few days. At the end of 10 days the arm was completely clear from shoulder to wrist. . .
> Following this the right arm was treated in the same way and ten days later the legs and trunk were treated. The result was that the arms were 95% cleared, the back 90% cleared, the buttocks 60% cleared, the thighs 70% cleared and legs and feet were 50% cleared.

The important feature of this treatment is that the response involved complex cognitive processes, since on each occasion particular somatic regions were named for clearance. So, let me, at least in outline, try to explain how my theory could account for some aspects of case 27.

According to the theory in the hypnotic state there occurs a tighter coupling between the ordinary matter brain and the SMM(brain) than in the normal state. This, in turn, could lead to an abnormally high exchange of gravitons between ordinary matter brain and SMM(brain), as suggested on p. 41. If then, in the hypnotic state the hypnotist suggests to the hypnotized subject (S) that a particular region of his body should be cleared of the black horny layer, then S's cognitive machinery would first have to recognize the meaning of the hypnotic suggestion. Since, by hypothesis, cognitive transactions, including the deciphering of meanings contained in language, are performed by the SMM(brain), this requires that S's SMM(brain) becomes activated via gravitons by S's ordinary matter brain.

Following this, S's SMM(brain) is assumed to process the hypnotic instruction so as to select, again by graviton exchange, appropriate nervous pathways of the ordinary matter brain. According to section 2.1.1 most somatic cells, with the exception of certain kinds of somatic cells (e.g. blood cells), are assumed to be uniquely labelled. Hence, nervous pathways, leading from the central nervous system to the periphery would consist of sequences of uniquely labelled nerve cells, for afferent as well as efferent pathways. If, now the appropriately 'set' SMM(brain) triggers, via emitted gravitons, the activation of specific efferent pathways, leading from the ordinary matter brain to labelled cells of the skin which are to be cleared of black horny layer, then nerve impulses could lead to action of nerve cell endings on the appropriate skin cells (or, perhaps, skin-related cells). By mechanisms, currently unknown, this could lead to separation of the horny layer from the skin cells. (Possibly emission of quasi-trophic substances from the nerve endings or from the skin-related cells could be involved.) At any rate, the locating of the skin areas to be cleared presents no problem for the present model. The accurate, unique, developmental wiring-in of the nervous system ensures that efferent pathways, whether activating muscles or somatic components of the skin (e.g. sweat glands), are precisely labelled and make accurate contact with target areas and commence at precise central nervous loci (see Wassermann, 1986).

One conceivable reason why hypnosis might promote the reception of, and emission of, psi-signals might be the assumed increased facility for exchanging gravitons between the ordinary matter brain and the SMM(brain) in the hypnotic state. This assumption was also pivotal in my explanation of the preceding Case 27. I must now return to cases of clairvoyance, and discuss one more case.

Case 28

This case concerns an experiment conducted by Dr A. S. Wiltse MD (who figured already in **Case 5**, p. 58) and which was reported by Mrs Sidgwick (1891, p. 76). The experiment was carried out with a hypnotized subject, Wiltse reported that:

> A Mr Howard lived six miles from me. He had just built a large frame house; our [hypnotized] subject had never seen the house, although, I presume, she may have heard it talked of. Mr Howard had not been home for some days, and asked if Fannie [the hypnotized subject] should go there by travelling clairvoyance and see if all were well. She exclaimed at the size of the house, but railed at the ugliness of the front fence, saying that she would not have 'such an old torn-down' fence in front of so nice a house. 'Yes' said Howard

laughing, 'my wife has been worrying the life out of me about the fence and the front steps.' 'Oh,' interrupted Fannie, 'the steps are nice and new!' 'She is off there,' said Howard, 'the steps are worse than the fence.' 'Don't you see,' exclaimed Fannie impatiently, 'how new and nice the steps are? Humph,' (And she seemed absolutely disgusted judging by the tone.) 'I think that they are real nice.'

Changing the subject, Howard asked her how many windows were in the house. Almost instantly she gave the number (I think it was twenty six). Howard thought it was too many, but, upon careful counting found it exact.

From my house he [Howard] went directly home, and to his great surprise, found that during his absence his wife had employed a carpenter who had built new front steps, and they had been completed a day or two before Fannie gave her account.

This, again, can be interpreted as a case of travelling clairvoyance, by means of the mechanisms already invoked in **Cases 15-17, 24-25**, showing also, once more, that the class characteristics of such (spontaneous) cases surface repeatedly. Yet, notwithstanding the repeated occurrence of class characteristics in cases of 'travelling clairvoyance,' in 'Out of the Body experiences' (OBEs) and other types of psi-phenomena, still to be discussed, there remain hard-faced, intellectually often very conventional, academics whom no amount of evidence will convince of the genuineness of psi-phenomena, whether this evidence were to be vouched for by a dozen or more Nobel Laureates, by the President of the United States or the President of Russia.

Such sceptics believe that *a priori* psi-phenomena cannot occur, so that relevant evidence either does not have to be examined, or must invariably be based on errors of the evidence given. (See also section 1.11 for related comments.) I have met at least one narrow minded professor of psychology, who had little academic glory to his name, but who rejected radically psi-phenomena on the grounds just cited. And I met numerous, academically even less well endowed, people who held similar views. To try to convince such people that psi-phenomena exist, have repeatable class characteristics and can be fitted into a coherent mechanistic materialistic theory consistent with modern physics, is to attempt the impossible.

3.2 Interpretations of more case histories of spontaneous telepathy, etc.

In section 2.3.1 I suggested already a mechanism for telepathy. Conceivably telepathy could also occur between surviving SMM(brain)s of dead people and SMM(brain)s of the living. I shall discuss the possibility of SMM(brain) survival after death of the ordinary matter brain later in much more detail. Frederic W. H. Myers (1903, vol.2, p. 195), one of the founders of the (London) Society for Psychical Research, argued already that if telepathy could occur between surviving discarnate 'personalities' and living people, then living (ordinary matter) brains might not be essential for telepathy, although Myers, who was a mentalist, as far as I know, did not invoke a physical SMM(brain), as is done within my theory. In terms of the present mechanisms the redundancy of the ordinary matter brain for telepathy could mean the following. Suppose that after destruction of an ordinary matter brain, after death, by decay, or cremation, the SMM(brain), formerly bound to the ordinary matter brain, survives intact. That it survives intact at least for a considerable time (or, perhaps, indefinitely), could be because there are Shadow Matter constituents bound to that SMM(brain) which stabilize its structure and which survive. Then telepathic communications, via SMM(complem)s, could occur between the surviving SMM(brain) and the SMM(brain)s attached to living ordinary matter brains. Likewise, the present theory could envisage telepathic communications between two or more surviving SMM(brain)s via SMM(complem)s.

I have discussed already near-death experiences (NDEs) and concurrent 'out of the body experiences' (OBEs) and noted that 'hearing' during NDEs and OBEs seems, according to Moody (1976, p. 42 and p. 47) to be of a telepathic kind (see p. 43), so that we have here prima facie evidence for telepathy in otherwise, apparently, unconscious subjects (as far as their reaction to stimuli is concerned). Also the case of Canon Bourne and his daughters (**Case 1**) suggests that telepathy between the three participants in the simultaneous apparition seems a plausible explanation. Perhaps some sceptics could argue that the subjects of NDEs during OBEs could still hear, despite apparent unconsciousness, so that unconsciousness was not complete, and that telepathy, as regards hearing of what other people said need not be invoked. Yet, the subject of at least one of the OBEs reported that her 'hearing' of other people was unlike normal hearing (p. 43) and seemed to resemble 'mind reading'. The following case is, therefore, interesting in this context.

Case 29

This case is taken from Myers (1903, vol.1, p. 659) and was reported in January 1883 by Mr R. Fryer of Bath. He wrote:

A strange experience occurred in the autumn of the year 1879. A brother of mine had been from home for three or four days, when one afternoon, at half-past five (as nearly as possible), I was astonished to hear my name called out very distinctly. I so clearly recognized my brother's voice that I looked all over the house for him; but not finding him, and indeed knowing that he must be a distance of some forty miles, I ended by attributing the incident to a fancied delusion, and thought no more about the matter. On my brother's arrival home, however, on the sixth day, he remarked amongst other things that he had narrowly escaped an ugly accident. It appeared that, whilst getting out from a railway carriage, he missed his footing, and fell along the platform; by putting out his hands quickly he broke the fall, and only suffered a severe shaking. 'Curiously enough' he said, 'when I found myself falling I called out your name.' This did not strike me for a moment, but on asking him during what part of the day this happened, he gave me the time, which I found corresponded exactly with the moment I heard myself called.

This case brings us back to Moody's reported OBEs, where the subject reported also hearing the complete speech of others whom they could see in their OBE. The explanation then was that what the subject 'heard' was telepathically transmitted from the SMM(brain) of other people. Likewise, **Case 29** suggests that there occurred telepathy between the SMM(brain) of the falling brother and the SMM(brain) of Mr Fryer, via SMM(complem)s. In fact, Case 29 makes the telepathic 'hearing' by subjects of OBEs a very plausible possibility.

Case 30

The following case, taken from the 'Report on the Census of Hallucinations' (1894) in the *Proceedings of the Society for Psychical Research* (London) vol.10 was contributed by Miss A.E.R. and written down in 1890. She wrote:

When out in camp in an Indian jungle, my sister and I were anxiously awaiting the return of her husband, who had left in the morning on a surveying expedition, promising to return early in the afternoon. Between 6 and 7pm we were uneasy, and were watching the line of road, I should say, 200 yards distant from where we stood. Simultaneously we exclaimed 'There he is' and distinctly saw him, sitting in his dog-cart driving his grey horse, the scythe occupying the seat behind. We at once returned to the tents - my sister ordering the bearer to get the Sahib's bathwater ready, and the butler to prepare dinner - I running to set my brother-in-law's mother's mind at rest as to the safety of her son. However, as time passed on, and he did not appear, our alarm returned, and was not allayed until he arrived in safety at eight o'clock. On interrogating him, we found that he was just starting from the surveying ground about eight miles distant, at the very time we had the above related experience. I should add that we were both in good health and certainly awake at the time, and I have never before or since had any experience of the kind.

Here, as in the Wilmot case (**Case 15**) and in **Case 1**, we have a simultaneously and collectively experienced apparition of a living person, and the simultaneity again required synchronization (as in Case 1). The telepathic hypothesis, which assumes telepathy between the two sisters, combined, perhaps, with clairvoyant perception of the husband by one of the sisters, could provide a possible explanation. The mechanisms in the two cases (concerning Mr Wilmot's and Mr Tait's simultaneous recognition of Mrs Wilmot's apparition in **Case 15**, and the simultaneous recognition of the husband of one by the two sisters in **Case 30**) could be the same. Also in **Case 1** there seemed to occur simultaneous telepathic synchronization involving three percipients of the experience.

There occur also 'auto-apparitions' in which people see themselves. This, as we saw earlier, happens typically in OBEs. One such auto-apparition is reported in the *Proceedings of the Society for Psychical Research* (London), vol.10, p. 74 in the 'Report on the Census of Hallucinations.' I have explained already how one's own body can be perceived in an OBE. Other auto-apparitions could function as follows. The SMM(body) of the percipient could emit SMM(complem)s which, in turn, could emit 'sphotons' (i.e. Shadow Matter photons). The latter could then act on the Shadow Matter eyes of the SMM(body) of the percipient (i.e. we have a feed-back reaction), and via these could have an effect on the SMM(brain) of the percipient, which then synthesizes the apparition. (Alternatively, the SMM(complem) could have interacted directly with SMM(brain) of the percipients.)

Case 31

The case was reported by Miss A.B.O., who wrote:

In June, 1889, about 8-9pm it being quite light at the time, I saw near - in Scotland, a figure approaching me, which, on coming near, I discovered was the double of myself, except that the figure, which wore a white dress, had a charming smile. I also wore a white dress; the figure had black on its hand, whether gloves or mittens I do not know. I had neither. It was out of doors, coming down the garden walk. On holding out my hand to it, the figure vanished. I was 24 years old, in robust health, and not in anxiety or grief at the time.

It could hardly be argued that the auto-apparition was caused by hypnotic suggestion, although we know that hypnotic trance can be used to induce, post-hypnotically, most intricate apparitions that are perfectly life-like (see **Case 3a** p. 42). Nor does the case bear any resemblance to **Case 6** (p. 51) where during an OBE a lady saw an apparition of herself, but that apparition shared none of the self-awareness of the lady, but knew about her 'problematics.' In the present case we are, as far as I can judge, not dealing with an OBE and the percipient did not claim to be out of her body. The possible mechanism was described above and may have involved also a measure of hallucination (e.g. the apparently hallucinated black gloves or mittens on the hands of the apparition, although this, like the apparition's smile etc., could be brought about simply by the constructive or creative cognitive ability of the percipient's SMM(brain); that such powerful constructive ability exists is clearly shown by the behaviour of the post-hypnotically induced 'apparition' in Gindes' experiment (see Case 3a p. 42). I have included this case here, because it provides an example which seems to rule out telepathy as an explanation in this instance.

Case 32

The following case from Gurney *et al.* (1886), also reported by Myers (1903, vol.1, pp. 657-58), was brought to notice by Major W. It seemed to involve telepathy between four people. Major W wrote on 9th February 1882:

It was the month of August; rather a dark night and very still; the hour midnight; when before retiring for the night I went, as is often my custom, to the front door to look at the weather. When standing for a moment on the step, I saw, coming round a turn in the drive, a large closed carriage and a pair of horses, with two men on the box. It passed the front of the house; and was going at a rapid rate towards the path which leads to the stream running, at that point, between rather steep banks. There is no carriage-road on that side of the house, and I shouted to the driver to stop, as if he went on, he must undoubtedly come to grief. The carriage stopped abruptly when it came to the running water, turned, and in doing so, drove over the lawn. I got up to it; and by this time my son had joined me with a lantern. Neither of the men on the box had spoken, and there was no sound from the inside of the carriage. My son looked in, and all he could discern was a stiff-looking figure sitting up in a corner, and draped, apparently, from head to foot in white. The absolute silence of the men outside was mysterious and the white figure inside, apparently of a female, not being alarmed or showing any sign of life was strange. Men, carriage and horses were unknown to me, although I know the country well. The carriage continued its way across the lawn, turning up a road which led past the stables, and so into the drive again and away. We could see no traces of it the next morning - no marks of wheels or horse's feet on the soft grass or gravel road, and we never again heard of the carriage or its occupant, though I caused careful inquiries to be made the following day. I may mention that my wife and daughter also saw the carriage, being attracted to the window by my shout. This happened on 23rd August 1878.

On the face of it one cannot rule out that there was a real carriage. Major W did not report whether the carriage was noiseless, or whether he heard it moving and could hear the horses move, or could feel the carriage when trying to touch it (i.e. he does not state whether he tried to touch it). If the case was one of an apparition, then it could be similar to **Case 1** where also several percipients participated and simultaneously 'saw' the apparition. As in Case 1, the explanation of the present case could be that one of the percipients' SMM(brain) synthesized the apparition, much as, presumably, the SMM(brain) of Gindes' hypnotized subject synthesized post-hypnotically, as a result of hypnotic suggestion, the complexly behaving 'apparition' of her dead brother (see p. 42). In the present **Case 32**, as in Case 1, the 'apparition' while modelled by the SMM(brain) of one of the percipients could then, via SMM(complem)s initiate the modelling of matching representations of apparitions by the SMM(brain)s of the other three percipients. As in Case 1, each percipient of this, possibly telepathically produced, multiple apparition would have to synthesize a similar apparition, but appropriately modified, so as to fit his or her position in the environment relative to the location of the apparition.

Gindes' case (Case 3a, p. 42) shows clearly the human capacity to synthesize immensely complex and intricately behaving apparitions which fit naturally into the existing environment, so that the syntheses of such highly complex apparitions in Case 32, and their apparent natural behaviour, as seen from the viewpoint of each percipient, is not surprising. Moreover, if, as much of the evidence suggests, there was no real carriage in Case 32, then the similarity of this case with Case 1 indicates that there exist in both cases repeatable class characteristics of this class of cases. This case, like Case 1, illustrates also again that cases which, presumably, involve psi-phenomena are spontaneous and sporadic, so that occurrence of such phenomena on demand (say on stage by a conjurer like James Randi (1991)), as is often requested by supposedly 'scientifically-minded' parapsychologists and others, seems unrealistic. (This is illustrated also by the many other case histories cited in detail above and below). Because of the urge of many parapsychologists to concentrate on the search for 'repeatable' experiments, which, such as the guessing of randomized cards, were conceivably 'repeatable to order,' although this was not really the case, less attention was paid to spontaneous cases, this century than last century, soon after the formation of the Society for Psychical Research in London. It is for this reason that most reported (or printed) cases of spontaneous psi-phenomena date back 90-110 years. This does not mean that such phenomena are not reported nowadays, but, because of misguided views concerning repeatability (e.g. disregarding repeatability of class characteristics of spontaneous cases) it has become fashionable to discard spontaneous cases, by and large, as non-scientific and not to present them in print in full detail, with occasional exceptions.

This, of course, as I have argued, is putting the cart before the horse. I can only repeat that the fact that spontaneous case reports have anecdotal character does not make these cases less significant for discovering the mechanisms of psi-phenomena than the quasi-anecdotal character of many past medical case histories is for understanding some aspects of particular, classifiable, diseases, and getting at the mechanisms of these illnesses. I think that it is true to say that if scientists had followed in the footsteps of James Randi in their attitudes towards exploring basic phenomena, then many important discoveries that were made would not have been made.

If I am right and, unpredictably, members of the public at large, provide the case histories of spontaneous phenomena of parapsychology, just as they provide the bulk of medical case histories (for human populations), then much (but not all) of the systematic search for psi-phenomena in a small number of laboratories which try to produce systematically phenomena that occur only as a rule spontaneously and sporadically may be misguided.

Nevertheless, as shown in **Cases 16, 24-5**, sometimes a subject in mediumistic trance or in hypnotic trance can be directed to 'visit' telepathically and/or clairvoyantly particular distant localities and report what is going on there. Hence, one must not jump to the dangerous conclusion that any form of experimentation in parapsychology is impossible, even if many of the most significant case histories, and their class characteristics have occurred spontaneously. As regards the production of psi-phenomena 'to order,' I think that it is of the utmost importance to remain open-minded, since possibly drugs, or other physical methods yet to be discovered, might in future greatly facilitate the occurrence of psi-phenomena. At present, however, although the kinds of experiments just mentioned (in Cases 16, 24-5) are occasionally very successful, the type of experiments involving card guessing or kindred techniques have so far not been impressive, with some notable exceptions (e.g. some experimental series by Dr Gertrud Schmeidler), and even these have been the subject of critical attacks (although Dr Schmeidler has rebutted most or all of these). In any case, I think that the intricate structuring of spontaneous case histories, which tell us that something quite specific has happened which has repeatable class characteristics, may be more important than a host of varied excuses why class characteristics of alleged psi-phenomena in the laboratory have not turned up repeatably. Perhaps many people are more interested in, possibly plausible, explanations why long experimental series with card guessing (etc.) do not, with the same subject, show up the same class characteristics indefinitely, than in observing repeatable class characteristics in spontaneous cases.

The frequent failure to capture on a significant scale in the laboratory, under non-spontaneous (i.e. experimentally controlled) conditions, psi-phenomena which may, for all we know (but not necessarily always in the future), only occur spontaneously 'in the wild' (but see Cases 16, 24-5 for notable exceptions to this statement), has encouraged certain experimental psychologists, and others (e.g. some magicians), to claim that psi-phenomena do not exist. This, of course, is patently absurd, and the same psychologists, not surprisingly (owing to their prejudice), also try, on ever-varying premises to explain away the host of known case histories of spontaneous cases in parapsychology, of which I have reported here only a small sample. These psychologists simply empty the baby with the bathwater. Their influence has, I believe, gone far beyond the merits of their case, because parapsychologists had no theory to offer which could unite psi-phenomena within a single theoretical framework consistent with physics. This is no longer the case. It was argued by some of these psychologists; and others, that if psi-phenomena are genuine then they would be 'contradicted by the laws of physics,' so that psi-phenomena could not be genuine (see also p. 114 and section 1.10). This type of ideology was, of course, based on the preposterous view that we necessarily know already all the laws and possible theories of physics. This view, however, has not been accepted in this book (see Introduction and section 1.1).

After this detour I shall return to more case histories and their interpretations in terms of my theory.

Case 33

The following case from Gurney *et al.* (1886, vol.2, p. 35), cited by Myers (1903, vol.1, p. 665) suggests that telepathy and clairvoyance were involved. The account is from Mr Richard Searle, barrister, who wrote on 2nd November, 1883:

One afternoon, a few years ago, I was sitting in my chambers in the Temple, working at some papers. My desk is between the fireplace and one of the windows; the window being two or three yards on the left side of my chair, and looking out into the Temple. Suddenly I became aware that I was looking at the bottom window pane, which was about on a level with my eyes, and there I saw the figure of the head and face of my wife, in a reclining position, with the eyes closed and the face quite white and bloodless, as if she were dead.

I pulled myself together, and got up and looked out of the window, where I saw nothing but the house opposite, and I came to the conclusion that I had been drowsy and had fallen asleep, and, after taking a few turns about the room to rouse myself, I sat down again to my work and thought no more of the matter. I went home at my usual time that evening, and whilst my wife and I were at dinner, she told me that she had lunched with a friend who lived in Gloucester Gardens, and that she had taken with her a little child, one of her nieces, who was staying with us; but during lunch, or just after it, the child had fallen and slightly cut her face so that blood came. After telling the story, my wife added that she was so alarmed when she saw the blood on the child's face that she fainted. What I had seen in the window then occurred to my mind, and I asked her what time it was when it happened. She said as far as she remembered, it must have been a few minutes after 2 o'clock. This was the time, as nearly as I could calculate, not having looked at my watch, when I saw the figure in the window-pane.

Mr Searle added that this was the only occasion on which he had known his wife to have a fainting-fit. She was in bad health at the time. To explain **Case 33** in accordance with my theory, one can assume that Mrs Searle's SMM(body) (including her SMM (brain)) could directly or indirectly generate SMM(complem)s corresponding to that SMM(body), some of which reached Mr Searle's SMM(Brain). Normally

Mr Searle's SMM(brain) would not respond to this kind of input, since, as I postulated earlier, response to the multitude and great variety of SMM(complem)s from the most varied sources (and involving also SMM(complem) replications) is unlikely in the normal state of tight binding between SMM(brain) and the ordinary matter brain (of Mr Searle and other humans). Yet, just as our normal responses to environmental stimuli are highly selective (owing, as is here assumed, to the selectivity of the SMM(brain)) so it can be assumed that Mr Searle's SMM(brain) responded selectively to that part of the SMM(complem) which represented Mrs Searle's SMM(brain) and its Shadow Matter representation of discomfort or distress. This could have aroused appropriate states in Mr Searle's SMM(brain). This telepathic mechanism, which could have functioned during a short spell of weakened bonding of Mr Searle's SMM(brain) to his ordinary matter brain, could, in turn, have activated Mr Searle's SMM(brain), so that it registered a SMM(complem) of the SMM(body) (including its clothes etc.) of his wife while in the fainting state, or just after, so that we could have a combined case of telepathy and clairvoyance. Mr Searle's SMM(brain) could then have combined the apparition of Mrs Searle with SMM(brain) representation of the environment i.e. the window-pane.

The following case is essentially similar to **Case 33** and can be explained similarly, demonstrating, once more, that different spontaneous cases of the same class may exhibit repeatable class characteristics, which, of course, allows us to allocate such cases to classes which have several members per class. (The class of apparitions concerned are 'crisis apparitions'.)

Case 34

This case comes also from Gurney *et al.* (1886, vol.2, p. 37) and concerns an apparition of a dying person. The case was reported by Mrs Taunton of Birmingham on January 15th 1884 and is also printed by Myers (1903, vol.1, p. 666). Mrs Taunton reports:

> On Thursday evening 14th November 1867, I was sitting in the Birmingham Town Hall with my husband at a concert when there came over me the icy chill which usually accompanies these occurrences. Almost immediately I saw with perfect distinctness between myself and the orchestra, my uncle, MW, lying in bed with an appealing look on his face, like one dying. I had not heard anything of him for several months, and had no reason to think he was ill. The appearance was not transparent or filmy, but perfectly solid-looking; and yet I could somehow see the orchestra, not through but behind it. I did not try turning my eyes to see whether the figure moved with them, but looked at it with a fascinating expression that made my husband ask if I was ill. I asked him not to speak to me for a minute or two; the vision gradually disappeared, and I told my husband, after the concert was over, what I had seen. A letter came shortly after telling me of my uncle's death. He died at exactly the time when I saw the vision.

Another group of cases involving, among other features, telepathy as a class characteristic are collective hallucinations. Some of these are discussed in the 'Report on the Census of Hallucinations' in the *Proceedings of the Society for Psychical*

Research (London) 1894, vol.10, pp, 25-422, Chapter 15. I have cited already some of the case histories from this report in order to explain them within my theory. Typical examples occur in Cases 1,15, 30 and 32. In **Case 1** the three percipients of the hallucination saw simultaneously the apparition of Canon Bourne (p. 24), while in **Case 15** (p. 97-8) Mr Tait and Mr Wilmot became aware simultaneously of the apparition of Mrs Wilmot, and in Case 30 (p. 116) Miss A.E.R, and her sister saw simultaneously the apparition of the sister's husband. I shall cite now a few more case histories of collective apparitions, because such case histories are not readily available to the public but are hidden away in the relatively few libraries or private collections that contain the appropriate volumes of the *Proceedings* or *Journals of the Society of Psychical Research* (London) or of other societies specializing in this field, or the volume by Tyrrell (1953) etc. In each case the explanation could be of SMM(brain) genesis of the apparition by one of the percipients, which then, by telepathy, generated a corresponding apparition in the SMM(brain)s of the other percipients. This demands, of course, that the SMM(brain) of each percipient constructs the percipient-specific apparition so as to fit naturally into the environment as perceived by the percipient. That ordinary matter brains (presumably jointly with SMM(brain)s, by hypothesis) are, indeed, capable of such adaptive construction of apparitions was seen earlier when I quoted Gindes' description of a hypnotically suggested, post-hypnotically appearing apparition of the dead brother of the subject (see Case 3a, p. 42). Here too the dead brother's apparition behaved perfectly normally and fitted naturally into the perceived environment.

Case 35

This case, taken from the 'Report on the Census of Hallucinations' (see p. 121) was reported by Mrs Greiffenberg and Miss Erni Greiffenberg to Mr F. C. S. Schiller, and headed 14th December 1890.

> In the beginning of the summer of 1884 we were sitting at dinner at home as usual, in the middle of the day. In the midst of the conversation I noticed my mother suddenly looking down at something beneath the table. I inquired whether she had dropped anything and received the answer, 'no, but I wonder how that cat can have got into the room?' Looking underneath the table, I was surprised to see a large white Angora cat beside my mother's chair. We both got up, and I opened the door to let the cat out. She marched round the table, went noiselessly out of the door, and when about half way down the passage turned round and faced us. For a short time she regularly stared at us with her green eyes, and then dissolved away, like a mist, under our eyes.

> Even apart from the mode of her disappearance we felt convinced that the cat could not have been a real one, as we neither had one of our own, nor knew of any that would answer the description in the place, and so this appearance made an unpleasant impression upon us.

> This impression was, however, greatly enhanced by what happened in the following year 1885, when we were staying in Leipzig with my married sister (the daughter of Mrs. Greiffenberg). We had come home one afternoon from a walk, when, on opening the door of the flat, we were met in the hall by the

same white cat. It proceeded down the passage in front of us, and looked at us with the same melancholy gaze. When it got to the door of the cellar (which was locked), it again dissolved into nothing.

On this occasion also it was first seen by my mother, and we were both impressed by the uncanny and gruesome character of the appearance. In this case, also, the cat could not have been a real one, as there was no such cat in the neighbourhood.

(An interesting, somewhat atypical, aspect of this case is that the same kind of apparition surfaced repeatedly on different occasions, in different surroundings, and was seen on both occasions by the same percipients, much of the time simultaneously. Also, on both occasions the cat dissolved into nothing, for both percipients, and before doing so seemed to fit naturally into the environment as seen by both percipients (compare this with **Case 32** and **3a**). Also obviously, **Case 35** like Case 32 is a collective apparition.

Case 36

This case, also from the 'Report on the Census of Hallucinations,' was reported by Dr S da G on the 17th March 1892.

I saw what seemed to be my fiancée's sister at a window in the garden. Her head was tied up in a handkerchief; I approached her, but on arriving opposite the window I found it closed. Nobody was there, yet one moment before I had seen the form, and I did not hear the window close, for which, indeed there was no time. I stood before the window, gazing at it in perplexity, when suddenly the panes seemed to disappear and the same form was leaning on the sill looking at me. It was not the sister of my fiancée; I recognized the appearance as that of my fiancée's mother, for I had seen her portrait in the house. I retired towards the place where my fiancée was sitting. I was horrified, but not to alarm her I did not run. When I came back to her she saw the form accompany me. It was visible only down to the waist. She had also seen what she supposed to be her sister at the window, and told me not to pay her any attention. There was at that time some misunderstanding between them. As I was going up the step of the veranda I felt as if there were a finger pulling me back by the collar. I did not look back, but G screamed out, 'Look - my mother!' and fainted away. Place Rio de Janeiro. Date 1876, hour nine o'clock pm.

At the time of the event, the mother had been dead for seven years. The light by which the apparition was seen was that of a gas-lamp which stood just opposite the garden gate. Just as in Gindes' hypnotically induced, post-hypnotic, apparition of a dead person (**Case 3a**, p. 42), the synthesis of an apparition of somebody deceased lies within the capacity of normal faculties of the ordinary matter brain (assumed to be acting jointly with the SMM(brain)). Although the percipient had only seen a picture of the dead mother, this could have sufficed for the brain-produced synthesis of the apparition. This apparition could then, telepathically, have invoked a corresponding (but differently adapted) apparition in the SMM(brain) of the fiancée. Also Dr S da G experienced a tactile hallucination. This, per se, is no more extraordinary than a visual or auditory hallucination.

Case 37

The following is a collective auditory hallucination taken from the *Proceedings of the Society for Psychical Research* (London) vol.10, p. 315. It involved Messrs de B and V of Rio de Janeiro and was dated 3rd and 4th of April 1892. Mr de B wrote:

I heard a voice call a friend who was with me in the same room, by his name 'Señor V,' Place Rua de S. Christina, Rio de Janeiro. We were both in bed in the same room. This happened in 1872.

I was still in bed, but awake. Health good. I naturally sympathized with ny friend, who had lost his wife the previous day. My age at the time was 42.

I did not myself recognize the voice; but it was the name used by my friend's wife in addressing her husband. The friend above referred to, Sr FV, also heard the voice. Each of us asked at the same time if the other had heard it. I was staying at his house to keep him company in his bereavement.

Although it is known that auditory hallucinations are symptoms of certain forms of schizophrenia, it could hardly be assumed that both percipients in this case were schizophrenics who had the same, very specific, auditory hallucinations at the same time.

There can be little doubt that the total number of paranormal hallucinations, here under discussion, has not significantly diminished since the census (by Sidgewick *et al.* 1894 in vol.10 of the *Proceedings of the Society for Psychical Research* (London)), and a more recent census organized by West (1948), a psychiatrist and a former professor of criminology at Cambridge University and repeatedly president of the Society for Psychical Research (London), shows this clearly (see West, 1954, p. 29-30). Yet, in order to assess whether the repeatable class characteristics of many of the older reported hallucinations also turned up in some of the case histories of the more recent survey, one would have to analyze the latter case histories in detail. Possibly West thought that the class characteristics most commonly found were already known from the older census (see Tyrrell, 1953), so that no further detailed reports and analyses of more recent case histories were required, particularly at a time (in the 1940s) when many parapsychologists were more interested in card guessing experiments (which in the supposedly most important case turned out to have been futile (see Markwick, 1978). Apart from this some people had little faith in the correctness of the accounts given by narrators of spontaneous cases. I must therefore turn to this topic as related partly to West's (1954) discussion.

3.3 How trustworthy are case histories of spontaneous cases?

Sceptics argue often that those, like myself, who take case histories of spontaneous cases seriously are foolish and naive and, supposedly, do not know the pitfalls of most, if not all, the hundreds of these cases, which in their opinion, are, perhaps, best forgotten about, or may serve as horrible reminders of how parapsychology can corrupt the minds of the innocent. So let me look at some of these arguments.

Before I do so, however, let me side-track a little into philosophy of science, a field in which I have worked and published over a period of years. It has been known for several generations, and, at least, well known through the writing of Sir Karl Popper, that science, both on the experimental and theoretical sides, is crowded with hypotheses. Now, the important thing about hypotheses is that, like the axioms of mathematics, they cannot be proved to be true (else they would not be hypotheses or axioms, respectively, but truths). Accordingly most, if not all, of science is composed of a network of unprovable hypotheses. This applies even to statements of facts. One must not confuse a fact with a statement about the fact. There lie many brain transactions between the observation of a fact, say the moon in the sky, and the statement about what one has seen or observed. We know perfectly well that the brain machinery that leads to the visual constancies in visual perception transforms what is presented to the retina into something that is often different. Similarly, transformations of what we factually perceive occur in all sensory modalities (such as sight, hearing, etc.) by brain mediation (whether due to an ordinary matter brain or a SMM(brain) or both combined). To overcome such problematics, scientists do most of their 'observing' not by means of their senses, but by means of complex observing machinery and recording apparatus. Yet, ultimately, at some stage, the machine-produced result of an observation has to be inspected by one or more human beings. The trouble here is that, like human brains, observing machines can produce artifacts of observations (notably in biology, but also in physics). Hence what we observe could always be an artifact of observation, Even if we observe it frequently. Scientists, therefore, try always to refine their observations in order to eliminate artifacts or observational errors as far as possible. Yet, total elimination of observational error is, in principle, not possible for reasons well known to scientists. So if observational error exists to a greater or lesser degree in all sciences, why should we expect it to be absent from observational parapsychology? Perhaps one safeguard against the influence of observational error is repeatability of phenomena.

Yet, even what is repeatable in physics does not amount to precise replication of data (because of variations in experimental error, due to ultimate fluctuations in the recording machinery etc.). No amount of improvement of measuring instruments can ever get rid completely of residual experimental errors. (The explanation for this can be found in the subject known as statistical mechanics, which deals with random fluctuations that cause error, among many of its other, more central, topics.)

Now, the human ordinary matter brain (and, presumably, also the hypothesized SMM(brain)) represent immensely intricate observation systems, when used in perceptions, and, also, when reporting hallucinations. It is, therefore, hardly to be expected that such systems should be error-free, when far less intricate systems, such as some of the most complex observing machinery used by physicists, are never completely error-free (however much error has been eliminated by careful design). To overcome the presence of error in physics, chemistry, molecular biology, etc., one can repeat an experiment several times and study whether the outcome remains approximately the same, within statistically acceptable limits of error (e.g. statistical 'limits of confidence.')

Similarly, in parapsychology, where one cannot deliberately repeat a situation in a spontaneous case, one can still show in many cases, as I stressed repeatedly, that different case histories may have similar class characteristics i.e. they have repeatable similarities. I noted early on that Moody discerned already typical class characteristics of OBEs, and similar remarks apply to various classes of apparitions, which have discernible class characteristics, and this applies also to other classes of psi-phenomena with repeatable class characteristics. Indeed, much the same state of affairs may apply in various studies of human psychology, where, because of genetic variation and differences in environments during development of the organism, no two organisms may be the same and may show distinct performance variations.

Yet, when we study, say, constancies of brightness, size and shape in visual perception we will find that, despite considerable inter-individual variations, these phenomena show typical class characteristics, which are reported with much variation from individual to individual (and, perhaps, are occasionally absent in exceptional cases). To believe that the repeatability of psi-phenomena should be of the relatively narrowly varying kind found in most, or all, branches of physics, and to adopt this physics paradigm for parapsychology, instead of allowing for the wide variations of properties often found in observational human psychology as the more appropriate comparison-paradigm, seems to me misguided.

Many people, who paid no attention to repeatable class characteristics of spontaneous psi-phenomena, thought that all spontaneous case histories could be ruled out *a priori* on the grounds of distortion of the original evidence with time (i.e. some cases were only reported to the Society for Psychical Research many years after their occurrence, e.g. Cases 34-5), or errors of memory, etc.

Thus, West (1954, p. 31) wrote:

Direct proof of fabrication in individual cases is naturally hard to come by, but a good example of this kind is the story of Sir Edmund Hornby *Proceedings of the Society for Psychical Research* (London) (1884) vol.2, p. 180, in the first edition only).

Sir Edmund was chief Judge of the Supreme Consular Court of China and Japan. In his account, which was published by the Society for Psychical Research in 1884, he states that he had been in the habit of allowing a certain newspaper editor - who also acted as a reporter - to call at his house in the evening and collect his written judgments for the day, so that they could be printed in the next day's paper. One evening in 1876, he wrote out his judgments as usual and gave them to his butler before he went to bed in case the reporter should call for them at a late hour. During the night:

Sir Edmund was aroused by a tap on the door and in walked the reporter. Thinking the man had mistaken the room, Sir Edmund told him to go and get the judgment from the butler. Instead of doing so the reporter approached the foot of the bed. He looked deadly pale. In a polite yet desperately insistent manner he pleaded with Sir Edmund to give him verbally a summary of the judgments. After some delay, and in order to avoid a commotion that might disturb his wife, Sir Edmund complied. The reporter made some shorthand notes, thanked the judge and left the room. Sir Edmund looked at the clock. It was half past one. Lady Hornby woke up, thinking she had heard talking. Sir Edmund told her what had happened. The next morning Sir Edmund heard that the reporter had died in the night. He had been seen writing at a quarter to one. At half past one his wife found him dead. Beside his body was his shorthand notebook with the heading for the judgment the last item in it. Sir

Edmund ordered an inquest. The reporter had died of heart disease. It was absolutely impossible for him to have left home during the night, and the butler was sure no one could have got into Sir Edmund's house.

Professor West continues:

The investigators should have been wary of this case. Corroboration was lacking, and the story of an apparition engaging in lengthy conversation is so unusual as to be suspicious.

Indeed, this case shows that to rely on any evidence which does not fit into known classes of cases is risky although some case histories must be the first known in their class. Moreover, even before West's book was published in 1954, Gindes (1953) published his book with the case history of the young lady (**Case 3a** p. 42) who, as a result of hypnotic suggestion, talked, as repeatedly mentioned, post-hypnotically to the apparition of her dead brother, and had a reciprocal conversation with that apparition (which had been ordered to appear six months earlier by the hypnotist).

So what happened in the case of Sir Edmund is essentially within the range of human capacities (at least when hypnotically ordered to occur). This may have been the first case that this happened spontaneously.

West continued:

But the status of the witnesses impressed the investigators [of Sir Edmund's case]. They quoted the case and commented favourably upon it in an article in the *Nineteenth Century*. In a later issue of this magazine (Nov. 1884), Mr F. H. Balfour, writing from China, supplied some surprising information. The reporter in Sir Edmund's story was Mr Hugh Long, editor of the *Shanghai Courier*, who had died between 8am and 9am on Jan. 21st 1875. There were no judgments made on Jan. 20th 1875. Sir Edmund did not marry until three months after the reporter's death. On hearing these facts, which proved indisputable, Sir Edmund admitted that his memory must have played him an extraordinary trick. He would hardly have invented the story and allowed it to be published, knowing all the while that it was false and liable to be exposed. It is more likely that he did have a vision of the dead reporter, but that the coincidence in time with the death, and the other striking details were added to the occurrence bit by bit until after eight years his memory of the affair had become completely false. Falsifications are liable to occur in anyone's account of an emotional scene, especially when it took place long ago. The unsubstantiated testimony of a single person, however authoritative he may be, can never be adequate evidence for a psychic experience.

With due respect to Professor West (and an army of others who regularly cite the Hornby case), one cannot generalize sweepingly from one case, containing grotesque distortions of evidence, and lacking repeatable class characteriestics, to the claim that this may apply to most or all other cases, notably cases, like those cited in this book which contain repeatable class characteristics (e.g. OBEs, apparitions etc.). It remains a reasonable hypothesis that the case histories reported in this book (except the Hornby case) are valid in all major details. To make a sweeping empirical generalization, as Professor West does and accept it, would be like accepting the faulty generalization that the Piltdown forgery discredits all of anthropology, including human paleaontology.

It would be like saying that one false witness in a court (and such people exist) discredits, or casts doubt on, the evidence of all witnesses in all courts of law for all time. I do not know whether Professor West, as a criminologist, would subscribe to such a generalization, yet, he seems to be doing something dangerously similar in the case of uncorroborated testimony by many narrators of spontaneous parapsychological case histories, particularly those that exhibit repeatable class characteristics. One cannot disregard hundreds or thousands of OBEs (which have repeatable class characteristics) simply because some sceptic demands this.

In many cases the verdict of a jury may ultimately depend on the credibility of a single key witness. If such credibility were not accepted in many cases, then our whole system of justice in, say, criminal cases would break down. Yet, if we accept key witnesses as *bona fide*, then we should, on equal grounds, notwithstanding Judge Hornby, also accept the accounts of those who claimed to have experienced paranormal events, particularly if these exhibit repeatable class characteristics, unless there are compelling grounds, as in the Hornby case, to discard such cases.

This of course, does not mean that corroborative evidence, if available, is not welcome, or should not be obtained, if possible. Yet, I doubt whether the presence of even 500 witnesses in each of most of the cases I have narrated above, and will cite below, would in the least ruffle any of the die-hard sceptics who are adamant that psi-phenomena do not exist, and, hence, cannot occur. After all, similar die-hards denied, without batting an eye-lid, that hypnotic faculties exist (see section 1.4) and that fraternity was for long undaunted by the most compelling evidence for the existence of these faculties. I am, therefore, not writing this book in order to persuade the group of severely prejudiced (e.g. most or all members of CSICOP see p. 64) of the errors of their views, since I believe that nothing, however well witnessed and coroborated, will do. Rather, my aim is to address myself to reasonable people and suggest to them that many case histories of spontaneous cases of ostensible psi-phenomena even if not corroborated by independent witnesses, could make sense within an integrated mechanistic materialistic theory that is related to recent theoretical physics.

Notes to chapter 3

1. In particular, my theory assumes that, according to the scheme for clairvoyance on p. 87, the SMM(body)s of Messrs Wilmot and Tait, and the SMM(object)s of other objects in the stateroom interacted via sphotons (i.e. Shadow Matter photons) with the Shadow Matter 'eyes -representing' structures of SMM(complem, Mrs W), and, hence, with SMM(complem, brain, Mrs W), thereby inducing memory traces of the visual appearance of Messrs Tait and Wilmot and the objects in the stateroom in terms of SMM(complem, brain, Mrs W) structures.

2. Prominent abdomen

3. The whitish light and experience of floating are also typical class characteristics of some more recently reported OBEs (see Moody, 1976) (see p. 37-8).

4. Chief Justice Way is the gentleman who acts as deputy for his Excellency when absent from the colony (noted by A. W. Dobbie).

5. The assumed strengthened coupling between SMM(brain) and ordinary matter brain in hypnosis could be expected to promote an abnormally high rate of exchange of gravitons between SMM(brain) and ordinary matter brain, which could result in the hyperactivity of the SMM(brain).

4.
Walnut Creek and other Synchronicity Experiences and their Interpretation as Psi-Phenomena

4.1 Examples of Synchronicity Experiences (SYNEXs)

Synchronicity Experiences (SYNEXs) are highly structured meaningful symbolic relationships that occur within a short time interval of each other, in everyday situations. A not particularly striking example of a SYNEX was cited in Case 2 (p. 26). I shall now cite and interpret numerous SYNEXs, mainly taken from my own experience, although I believe that such SYNEXs are widely experienced by the public at large, and not just by myself. The difference between the 'public at large' and myself is that whereas I am interested in analyzing SYNEXs in terms of psi-phenomena, which, in turn, could be explained in terms of my unificatory theory of these phenomena, the 'public at large' generally ignores SYNEXs. If it regards them at all, it ascribes them to 'pure chance coincidences,' i.e. as something that needs no explanation. Yet, whether something needs an explanation lies in the eye of the beholder. To a five year old child, the observation that a magnet picks up iron filings does not seem to require a sophisticated explanation, whereas a theoretical physicist, specializing in ferro-magnetism, may invoke complex theories of atomic physics in order to explain such phenomena.

Case 38

On 15.4.1988 I was due to visit my late friend Michael Salinger and his family in Durham (England), and did so. The day before on 14.4.1988, I went to London by train. On the train sat a lady nearby who read J. D. Salinger's *The Catcher in the Rye*.

Here we have the same meaningful symbol 'Salinger,' which is not very common, occurring within two days, or less, of each other to the same percipient. I wish to suggest that SYNEXs, instead of being mere 'coincidences,' could provide some of the best indirect evidence for psi-phenomena. In principle, of course, any lawful behaviour of a physical system (e.g. a thin wire obeying Ohm's law) could be labelled a 'mere coincidence' for all measurements that conform to the law (within limits of error). That physicists try to attach deeper meanings to laws, and explain many of these laws as deductions from complex theories, is due to the fact that these physicists try to, and often can, produce a deep theoretical coherence among apparent 'mere coincidences'. It is, therefore, tempting to do the same as regards SYNEXs, and try to interpret some typical SYNEXs in terms of the present physicalistic theoretical framework for psi-phenomena.

In **Case 38** it could be suggested that my SMM(brain) had established a memory trace representation of my intended visit to the Salinger's, and this, by means of mechanisms already discussed, helped to make me aware clairvoyantly of the title page of the book carried by the lady on the train, and that, accordingly, I sat near her.

In my experience SYNEXs seem to occur clustered, with several SYNEXs coming over a span of several days, followed by a pause, which may be very long (i.e. weeks or months)(p. 27), during which no SYNEX is experienced by the same percipient. A possible explanation of this was given on p. 27. When SYNEXs come they are often very striking and intricate, as will be seen by cases of my own collection and examples cited by C.G. Jung (1955). (Perhaps one is apt to overlook less striking cases.) Jung (1955) believed that SYNEXs result often from an 'acausal connecting principle.' Such a principle, however, has no place in the physical materialistic sciences. In contrast to Jung, I shall argue that SYNEXs can be explained in terms of psi-phenomena (e.g. clairvoyance and telepathy), which, in turn, as I argued in much detail in the preceding chapters, could be interpreted in terms of a possible physics of Shadow Matter. If so, then by Occam's razor (i.e. the principle of economy of assumptions), the alleged 'acausal connecting principle' of Jung becomes redundant. In fact, if, as I contend, SYNEXs are explicable, in most or all cases, in terms of psi-phenomena, then SYNEXs may serve as strong *prima facie,* indirect, evidence for the existence of psi-phenomena.

Case 39

On 20th December 1983 at lunch time I addressed a Xmas card to my father's (half-Jewish) cousin Georg Beer. He lived then with his wife Edda at 1332 Canyonwood Court WALNUT CREEK (CA 94595) USA. When addressing the letter to Georg, while in town, I wondered why there should be a place with an extraordinary name such as Walnut Creek. That same day, as I returned to my house in the evening, I found two letters lying in the entrance to my hall (which had been delivered that morning after I had left home). Both letters were from Walnut Creek, California. (My home is in Newcastle upon Tyne, England). The first letter was from Georg Beer with Xmas and New Year wishes, the other letter was from old friends, Rose and Ernest S who also lived in Walnut Creek, California. The S's being Jewish (like myself) did not send a Xmas card, nor do they normally write specifically at Xmas time (they were friends of my parents in Germany).

Here, then, we have a typical SYNEX, where the same, often clustered, symbols, Walnut Creek in this case, turn up repeatedly (in this case three times) within a short span of time. Since the two letters had arrived at my house that morning (there was then only one mail delivery), they must have been lying in my hall at the same time when I wondered about the (to me) unusual name of the place called Walnut Creek (obviously not unusual to people in California). One possible explanation of this highly structured SYNEX is that it represents a case of pure clairvoyance (or what Puthoff and Targ (1981), Targ and Puthoff (1974) and Tart *et al* (1980) have called remote viewing). By pure clairvoyance, mediated by SMM(complem)s, my SMM(brain) could have perceived the envelopes and their contents, which contained the name Walnut Creek. This could have started my thoughts about the peculiarity of the name Walnut Creek and also made me write to Georg and Edda Beer.

I conclude that **Case 39** is prima facie possible evidence for clairvoyance. In **Case 2** (p. 26) it also looked as if SMM(brain)s could, say, via SMM(complem)s, receive clairvoyant signals emanating from the inside of closed books. If these interpretations are valid, they show, consistently, that people can obtain clairvoyant information about the interior contents of objects, such as letters, books, houses etc. This, of course, is in agreement with my general theory.

Case 40

This occurred in the 1950s. One evening I decided suddenly, for no particular reason, to visit the library of my university (the University of Newcastle upon Tyne, then part of the University of Durham). I went to the history section and decided to take out a biography of Napoleon by Kircheisen. On returning home, that same night, I read exclusively the account of Napoleon's stay and policy while confined to the Island of Elba. Next morning I received a letter with an invitation to attend a totally unexpected conference by some parapsychology group at the Palazzo Fonte de Napoleon located near Poggio on Elba.

Unless one wishes to invoke foreknowledge as an explanation,[1] the most economical explanation in terms of psi-phenomena of this highly structured SYNEX (which centres round the symbols Napoleon and Elba) is, perhaps, again to interpret it as a case of 'pure clairvoyance'. The letter was obviously in the mail at the time I went to the university library, and I could have perceived its contents clairvoyantly (just as in **Cases 2** and **39**) and this could have activated my SMM(brain) to select the book on Napoleon and to read the chapter about Napoleon on Elba. (Alternatively, telepathy between the organizer of the conference and myself is conceivable as an explanation.)

Case 41

The clustering of meaningful symbols, already noted in Case 39, is also well illustrated in the following case, reported by C. G. Jung (1955, p. 14). He wrote: I noted the following on April 1, 1949: Today is Friday. We have fish for lunch. Somebody happens to mention the custom of making an 'April Fish' [i.e. April Fool] of someone. That same morning I made a note to an inscription which reads: 'Est homo totus *piscis* ab imo.' In the afternoon a former patient of mine, whom I had not seen in months, showed me some extremely impressive pictures of fish which she had painted in the meantime. In the evening I was shown a piece of embroidery with fish-like sea-monsters in it. On the morning of April 2 another patient, whom I had not seen for many years, told me a dream in which she stood on a lake and saw a large fish that swam straight toward her and landed at her feet. I was at this time engaged on the study of the fish symbol in history. Only one of the persons mentioned here knew anything about it.

Having fish on a Friday is customary in many families for reasons of religious tradition. Since Jung was working on a study of the fish symbol in history, his, hypothesized, SMM(brain) could have been activated so as to give strong preference to the fish symbol (i.e. to create what psychologists call a 'set'). For instance, Jung could, by telepathy, have influenced the former patient so as to dream about the large fish (the telepathic process need not have led to awareness that it was going on, but could have been at at an unconscious level). Likewise, Jung's SMM(brain) could have acted telepathically (without he being consciously aware of this) on the person who mentioned the 'April fish'. Jung's Latin note is not surprising in view of his dominant attention to the fish symbol in history. Again, the patient who showed Jung 'some extremely impressive pictures which she had painted' could have been telepathically influenced by Jung's SMM(brain) and its directly, or indirectly, produced SMM(complem)s. The latter, by acting on that patient's SMM(brain) could have caused her to paint the pictures of fish, and to show them on that day to Jung. It must be stressed that in such telepathic exchanges neither the 'sender' nor the 'receiver' (of SMM(complem)s) need be consciously aware that such SMM(complem)s had been produced, transmitted and received, respectively. The fish-symbol-orientated activity of Jung's SMM(brain) could have peaked so that telepathic processes could have prompted the various people concerned to report to Jung all within the span of two days.

It did not occur to Jung to interpret SYNEXs in terms of psi-phenomena. Instead he invoked his 'acausal connecting principle' not just to 'explain' SYNEXs (if one wishes to call this an explanation), but also to explain psi-phenomena acausally (see Jung, 1955, p. 14). It is sometimes thought by laymen that quantum mechanics has abandoned causality. This, however, is not the case, since quantum mechanics retains (e.g. in the case of the time-dependent Schrödinger equation) a statistical determinism, so that events are statistically determined. Accordingly, quantum mechanics provides no pretext for introducing an 'acausal connecting principle.' Although Jung's (1955) essay formed the first part of a book written jointly with the eminent Nobel Laureate physicist Wolfgang Pauli (who wrote the second essay of the book), it is not obvious that Pauli saw any need for introducing an acausal connecting principle, even if he accepted some of Jung's other interesting ideas. While Pauli was interested in the early mystical roots of some scientific ideas, notably some of Kepler's ideas, this does not mean that Pauli welcomed back mystical principles into science. Views akin to Jung's 'acausal connecting principle' surface also in a book by Hardy, Harvie and Koestler (1973), who believed 'that a "coincidence-generating" acausal factor may operate in biological evolution and produce meaningful convergences, clusters, emergence of the new, and possible psi occurrences' (cited from Chari, 1977, p. 817). The postulated 'coincidence-generating' acausal factor seems to me to be as useful as the philosopher Bergson's (1911) *élan vital*, which was a metaphorical agent that was supposed to be behind evolutionary and other biological processes. Sir Julian Huxley (1942, p. 458) remarked that invocation of Bergson's *élan vital* in biology is comparable to 'explaining' the mode of propulsion of a railway engine in terms of an *élan locomotive*. I fear that the acausal factor of Jung and of Hardy, Harvie and Koestler belongs to the same (non-mechanistic) genre as Bergson's *élan vital*.[2] Beliefs in an acausal connecting principle strike me as inappropriate as some of the late J. B. Rhine's views (cited by Wolman, 1977b, p. 861). Rhine and Pratt (1957, p.10) maintained that psi-phenomena 'are distinguished from the other phenomena of psychology by the fact that they can be shown to be non-physical in character.' Obviously, I do not share the late Professor Rhine's point of view.

Case 42

On Sunday 12 April 1987 I went for a stroll to Grey Street, Newcastle upon Tyne. There in the shop window of the Waterstones's bookshop, I saw several copies of a book by Primo Levi, entitled *If Not Now, When?* with large adverts praising the novel. Up to that time I had not heard anything about Primo Levi (an Italian Jew, who survived Auschwitz), although he is a distinguished novelist, The following day, Monday 13 April 1987 I found the obituary for Primo Levi in *The Times* (London, p. 16). He had died on 11th April 1987, aged 67, having tragically committed suicide. *The Times* mentioned also Levi's novel *If Not Now, When?* (among other works).

So here we have a SYNEX with the symbols Primo Levi and 'If not Now, When' surfacing twice together to the same percipient within twenty four hours. It could have been 'pure clairvoyance' (i.e. remote vision) which led to my SMM(brain)'s pattern recognition of Primo Levi's obituary in *The Times*. This recognition process, in turn, could have interacted with my SMM(brain)'s memory traces of Primo Levi's book in the shop-window the previous day, and this attentive interaction could have led to my looking up the obituary, which I had perceived already clairvoyantly. Here, as in **Cases 2** and **39**, a clairvoyant process could have acted which revealed to the SMM(brain) of the percipient of a SYNEX information contained inside a book, or inside letters, or inside a newspaper. Of course, a sceptic could argue that all these SYNEXs are pure coincidence. Yet, the occurrence of so many of these highly structured coincidences would seem strange, particularly if clairvoyance, whose existence seems to be implicated independently by quite different types of psi-phenomena, could readily explain many of these SYNEXs. At any rate, the assumption that many SYNEXs are caused by the suggested psi-mechanisms of clairvoyance, seems to me preferable to the invocation of a non-scientific, metaphysical, acausal connecting principle. For my part I prefer to interpret symbol coincidences of SYNEXs as being caused by mechanisms, wherever possible, since this is part of the scientific attitude.

Case 43

I visited my second cousin Mrs B.S. in Birmingham from Monday 23rd March 1987 until Wednesday 25th March 1987. On 25th March she told me that she was going to Germany soon to meet a relative from Argentina, who could not come to England because of inability to obtain a visa (because of the recent

Falklands war). She told me that she would also like to spend some days in the Black Forest and asked me where to go. I told her that I liked Bad Dürrheim particularly with its dense wood, since I had been there twice as a boy in the Jewish Children's Home (the Friedrich Louisen Hospitz). On the following day, 26th March 1987, after my return to Newcastle, I received a letter (dated 18th March 1987) from Mr Gideon Bar-Joseph (a former official in the Israeli Knesset) who lives in Jerusalem, and who had not been in touch with me since I visited him in 1980. He wrote, among other things 'I wanted to ask you if you have any recollection of our common holiday trip in August 1929 to Bad Dürrheim in the Black Forest, where we spent some three weeks in the Jewish Children's Home there?'

So here we have the same symbols, Black Forest, Bad Dürrheim and Jewish Children's Home, surfacing twice, in totally different contexts for my (presumptive) SMM(brain) on two consecutive days. These symbols when combined are not of common occurrence. It could be argued again that my SMM(brain) perceived by pure clairvoyance the contents of the (sealed) letter from Jerusalem (and similar availability of the contents of sealed letters, closed books etc., by clairvoyance, was noted in *Case 42*), since the letter was already written and posted at the time Mrs S asked me about the Black Forest. This clairvoyant effect on my SMM(brain) could then have established, in terms of Shadow Matter hardware of the SMM(brain) an appropriate psychological set (or attitude) which made me answer Mrs S's question in the way I did. Possibly, in addition the (presumptive) SMM(brain) state caused by its Shadow Matter representation of my psychological set could have invoked, by telepathy, the thoughts about the Black Forest in Mrs S's SMM(brain).

Case 44

On 18th February 1987 I told Miss X that I had been misled by the clerical title of the Archbishop of York, Dr John Habgood. I had written a paper on philosophy (Wassermann, 1988a) in which I referred to his Grace's article in *The Times* (London) of 12th January, 1987 p. 12. This article was signed John Ebo. I was surprised that John Ebo, like John Habgood, seemed both acquainted with quantum theory, but thought at first that John Ebo was Dr Habgood's successor, although I could not find on the 18th of February any John Ebo in *Who's Who*. I asked a colleague, who informed me that John Ebo is His Grace Dr John Habgood. On 19th February Miss X rang me to draw my attention to a letter in *The Times* of 18th February 1987 p. 11, concerning titled clergy, by Vincent Strudwick, Director of the Oxford Diocesan Council of Education and Training. Canon Strudwick wrote that Anglican Bishops sign their names including a piece of territory of diocese they serve. He noted that in England a shorthand version in Latin [of the diocese] is used, and continues that his 'former Bishop (of Oxford) thus signed himself 'Patrick Oxon'.'

Here the SYNEX concerns a concept, namely the concept 'clerical signature'. Since I received *The Times* (London) regularly during the week, but had not read Canon Strudwick's letter (of 18th February) until Miss X drew my attention to it on 19th February, it could be that by pure clairvoyance, as in the preceding cases mentioned (**Cases 42-3**), my conjectured SMM(brain) perceived the Canon's letter in *The Times* on 18th February and that this led me to think about John Ebo and to look up *Who's Who* that day.

Case 45

During the week 6-12th May 1985 I asked Professor SP (a visitor to the School of Mathematics of the University of Newcastle upon Tyne) whether he intended to do any more work on the possible classical foundation of Fermi-Dirac Statistics in Statistical Mechanics. He told me that he and Professor PR had been using Grassmann Algebra in their work, and stressed that it differed drastically from Clifford Algebra (also a linear algebra). It had never occurred to me, since I don't work in that field, that Clifford Algebra and Grassmann Algebra could be related. SP told me that, according to their theories, one of these algebras would lead to mutually repelling particles, the other to mutually attracting particles. On 16th May, 1985, totally unexpectedly, I received a letter from Professor John Archibald Wheeler, a celebrated American theoretical physicist (at the University of Texas, Austin). We had never met or corresponded until then. He asked me to send him a reprint of my paper 'Quantum Mechanics and Consciousness' (Wassermann, 1983). This itself was a surprising request, and he explained later that he had seen a reference to my paper in the Vice-Chancellor's report of the University of Newcastle upon Tyne when he came to visit it (although we had not met). Wheeler enclosed with his letter two reprints, one entitled 'Bits, Quanta, Meaning' (reprinted from *Problems in Theoretical Physics* edited by A. Giovannini, F. Mancini and M. Marnlaro, University of Salerno Press, 1984). When I looked at the bibliography of Wheeler's paper I saw in note 3 a reference to a paper by E.R. Caianiello and A. Giovannini, titled 'Pure spinors as Pfaffians connecting Clifford and Grassmann algebras' (see *Letts. Nuovo Cim.* 34, 301-304 (1982)).

Thus, in this highly sophisticated SYNEX the symbols involved are connections between Clifford and Grassmann Algebras. The symbols Clifford Algebra and Grassmann Algebra turned up independently of each other (despite their relative rarity of use) within a few days from different sources, and apparently the Newcastle group and Professor SP did not know Giovannini's paper. Again, a possible explanation of this case could be similar to those already given in several of the other cases, in terms of pure clairvoyance. Professor SP's SMM(brain) could, by means of clairvoyance, have perceived the contents of the envelope of the letter (then already dispatched by Wheeler from the USA) and the title concerning the connection between the two algebras, which SP at that time did not know of. Likewise SP could by pure clairvoyance have perceived my typed name on the envelope that contained the reprints, and this could have motivated SP (unconsciously) to tell me about his views on the two algebras. It remains to explain why Professor Wheeler sent me reprints relating to the two algebras and their connections, although this is not a field that normally interests me. It could be argued that Professor Wheeler's SMM(brain) perceived telepathically signals from the SMM(brain)s of Professors SP and PR relating to their work on the algebras, and also received clairvoyant signals that

135

indicated that I was in close geographical proximity to the two professors. This could have motivated Wheeler (without him being consciously aware of what motivated him) to send me the reprint.

It should be noted that such complicated cases, which involve very subtle multi-person telepathy and/or clairvoyance, are also familiar to parapsychologists from the cross-correspondence cases which I shall discuss in chapter 5. What is remarkable is that here, in an everyday academic setting, there turned up evidence for psychic phenomena as complexly structured as some of the best cross-correspondences, namely evidence which seems to involve bits of data known to several (presumptive) SMM(brain)s, which despite the wide geographical separation of one (or in other cases more than one) of the SMM(brain)s are fitted together, like bits of a jigsaw puzzle.

Case 46

On December 7th 1989, in the afternoon, I visited the Wiener Library (London) and asked Mr A. Wells, one of the superintendents, whether he could supply me with the names of some prominent Jewish poets. He started his list, which he wrote down, with Osip Mandelshtam, an apparently famous Soviet-Jewish poet and writer, of whom I had never heard. Mandelshtam lived during the Stalin period of the 1920s and 1930s, and was murdered as a result of having written a satirical poem about Stalin. On the morning of the 9th December 1989 I travelled back on the train from London to Newcastle upon Tyne. Next to me sat a gentleman who was reading the *Independent* of 9.12.1989. The article he was reading was, as could be seen from the headline, about Mandelshtam and included a photograph of the poet. Somewhat later in the morning I asked the gentleman whether I could read the article (on p. 31), and he gave me the newspaper on loan.

Apart from possible precognition (foreknowledge) there could be an explanation in terms of telepathy in this case. I was at that time writing a paper, since published in *Philosophy* (1990, vol, 65, pp. 355-65), showing the many serious mistakes in Wittgenstein's grossly misguided onslaught on the Jewish people (he was himself three quarters Jewish). In the course of writing that paper I had come across allegations by various, notorious, anti-semites, that Jews could not write poetry (*pace* the German-Jewish poet Heinrich Heine, the Hebrew poet Nahman Bialik and Osip Mandelshtam etc.). I therefore concentrated on this topic, thus generating a psycho-logical set and, presumably, according to the present theory, a corresponding state of my SMM(brain). This could have led me by telepathy to look up Mr Wells at the Wiener Library (or, more likely, a combination of telepathy and clairvoyance could have been involved in leading me to Mr Wells) and by clairvoyance I could have become aware that the article on Mandelshtam was in the *Independent* carried by that gentleman, and, thus, seated myself next to him on the train in London.

Case 47

In 1984 I was a Visiting Professor at the Institute of Evolution of the University of Haifa (Director, Professor Eviatar Nevo). On Wednesday 14th November I gave an invited lecture to Professor Y. Fried's group at the Day Hospital and

Mental Health Clinic, Medical School of Tel-Aviv University, presenting a new biochemical and medical model of the possible causation of schizophrenia, which was later published in modified form (Wassermann, 1986a). During the lecture I mentioned a recent correspondence in the journal *Nature* (London), namely a letter by Dr Joshua Bierer (see *Nature* (London) (1983) 'No to Drugs' vol. 305, p. 468) whose views I did not share, and of whose existence I knew only because of his correspondence in *Nature*. After the lecture a lady whom I did not know came to me. She explained that she came originally from Germany, but had been associated with some psychiatric establishment in Britain and that she lived now in Israel. She asked me whether Bierer was still alive, since he must be in his 80s. She had known him. I told her that, judging by the recent correspondence in *Nature*, he might still be alive. On January 5th, 1985, after I returned to Newcastle, I started to read the front page of the German-Jewish refugee journal *AJR Information* (AJR= Association of Jewish Refugees) (vol.40, no. 1), a journal which I receive monthly. Although I had received the journal a day or two earlier, I had not had time to look at it, and had not glanced at the central pages. After reading as far as page two (and the pages are very thick and not transparent), I stopped for a while and began to think of my schizophrenia paper, and I thought also of the old German-Jewish psychiatrist, whom I had mentioned at Tel-Aviv, but could not quite remember his name, except that it started with Bi. Accordingly, I looked up the bibliography of my paper (read in Tel-Aviv) and found Bierer's name. After this I resumed reading *AJR Information,* and, about twenty-five minutes later glanced at page seven, where I noticed an obituary of Dr Joshua Bierer, who, at the age of 83, had died in Tenerife. (He was a pioneer in social psychiatry, and a very distinguished German Jew, who settled in Britain in 1938.)

This case (which is a SYNEX in its second part) is based on the symbol Bierer. The second part of the case could be explained in terms of pure clairvoyance. Before I had noticed Bierer's obituary on page seven, I had stopped several pages earlier to think about Bierer and to look up his name in my bibliography. This sudden thought about my schizophrenia paper and about Bierer could have been due to my (presumed) SMM(brain) perceiving clairvoyantly the obituary notice on page seven, which I had not opened or seen at that time. Thus, here, once more, as in several previous cases, we have evidence which is consistent with the hypothesis that SMM(brain)s could clairvoyantly (by mechanisms stated) perceive unknown contents of the inside of letters, books, or newspapers (etc.) not yet opened.

Case 48

I had typed out Case 47 for the first time on 26 October 1987 (from notes I had made earlier). The following day, 27th October 1987, on a journey to London, I read in *The Times* (London) p.161 the obituary of Wing Commander Jim Hallowes, who was a highly distinguished flier in World War II. The obituary notes that 'he died while on an extended holiday in Tenerife.' So, here, within twenty-four hours I came twice across the symbolism death in Tenerife, first

137

on 26th of October in connection with **Case 47**, and on the following day in connection with the present case.

Typing Case 47 could have produced a physical representation of a psychological set, in my presumed SMM(brain) for the symbolism, or concept, Death In Tenerife. This could have made my SMM(brain) receptive to the clairvoyantly perceived obituary in *The Times,* which then made me look at the obituary of the Wing Commander.

We have now encountered case after case, where widely different SYNEXs could be explained consistently in terms of either pure clairvoyance or telepathy, or telepathy combined with pure clairvoyance in which a SMM(brain) received information about printed (etc.) material contained in concealed form inside envelopes, books, newspapers (etc,) quite often far removed from the percipient. Undoubtedly, some people, as noted earlier, may prefer to consider SYNEXs as mere chance coincidences, whereas others, who try to detect causal connections in nature, might find the causal explanations given here more satisfying.

Case 49

On 28th August 1984 I went together with Miss W to the Edinburgh Festivals to hear, together, at 8pm that night in the Usher Hall, an all Mozart performance consisting of three works, the overture to the *Magic Flute,* followed by the *Jupiter Symphony* and, after an interval, by the *Coronation Mass.* On 29th September 1984 I told an acquaintance Mrs GH in London about the Mozart concert and its programme. On 1st October 1984 I saw (in Newcastle) in the *Radio Times* that they were televising a dramatized version of Graham Greene's novel *Doctor Fischer of Geneva,* and decided to watch it. Dr Fischer's daughter Anna-Luise stated that she was very fond of Mozart. After she got married to Mr Jones, Jones went to a record shop and asked for a copy of the *Jupiter Symphony.* The shop assistant said that there was a very good recording by the Vienna Philharmonic Orchestra, but that this was out of stock. So, Anna-Luise, who was with Jones, asked the assistant whether she could have Mozart's *Coronation Mass* instead.

So here we have a SYNEX which links the symbols Mozart's *Jupiter Symphony* and Mozart's *Coronation Mass*, and both symbols surface in conjunction first on 29th September 1984 and, again, on 1st October 1984 in totally different contexts for the same percipient (i.e. myself). It could be argued that the BBC programme producer, and many other people who had read Graham Greene's novel (but I had not) had formed memory traces of the conjunction of the two Mozart works, and that I became telepathically aware of this. My previous listening to the two works at Edinburgh and my talking about them to SH could then have formed a psychological set in my presumed SMM(brain) for these works, and the telepathic information that these works played jointly a part in Graham Greene's novel, and clairvoyant information from the *Radio Times* that the play was on, could then have prompted me to look at it. There are, however, alternative ways of explaining this SYNEX in terms of psychic phenomena (and similar remarks apply to the explanations given already of various other SYNEXs).

Case 50

On Sunday 16th September 1984, I listened to a BBC Radio 3 broadcast entitled 'Evolving Ideas' in which Professor Stephen Jay Gould of Harvard University was interviewed by Colin Tudge. Early in his interview Gould stated that Darwin was guided by the idea that populations would increase rapidly unless they were limited by limiting events. I could not, just then, remember the name of the well known man, who had predicted this unlimited population increase. Next day, 17th September 1987, I looked at the 'letters to the Editor' columns in *The Times* (London) (which I inspect irregularly). There appeared that day a letter by Alexander Murray of University College Oxford, headed 'The Pope and Marxism.' Its opening sentences read 'Graham Greene (September 11) reminds us, *a propos* the Pope and Marxism, that Karl Marx disapproved of Henry VII's dissolution of monasteries. Marx also had words for the clergy with families who, in the person of *Malthus* (and some of his disciples), preached to the families of the poor.'

Here, then, was the missing name Malthus, which the day before I had tried to recall without success! One explanation, in terms of my theory, could be as follows. My SMM(brain) was presumably 'set,' to recall the name Malthus, and such a 'set' (i.e. its physical SMM(brain) representation) could last a long time. Moreover, by pure clairvoyance I could have perceived the letter in *The Times,* and my persistent 'set' for Malthus, combined with the clairvoyant information, could then have led my SMM(brain), jointly with my ordinary matter brain, to an inspection of the letter in *The Times.*

Case 51

The following case history, although involving remarkable symbol coincidences, is, perhaps, a somewhat atypical SYNEX, since it lacks the time-delay factor between repeated surfacing of the symbols.

On 28th August 1984 Miss W and I went to the Edinburgh Festivals (see also **Case 49**). In the morning at 11am we went to the Queens Hall, where we had booked seats C11 and C12 to hear Yo-Yo-Ma perform three unaccompanied cello suites by J.S. Bach. I sat in seat C12 and Miss W in seat C11 on my right. Two seats to my left, in the same row, sat a lady, who suddenly leaned over and asked Miss W 'aren't you Miss W?' The questioner was a Miss S, then resident as a scientist in Edinburgh. Miss S had been taught by Miss W in a school near Newcastle until 1961. Miss W then introduced me to Miss S, saying 'this is Dr Wassermann'. Whereupon Miss S replied 'I know him, he taught me Applied Mathematics at Level 11 at Newcastle University (then part of the University of Durham) during the year 1961-1962.' (I could only very vaguely remember Miss S, certainly not by name, since she had been in a very large class, most of whose members I did not know personally.)

This case could involve multi-party telepathy, not unlike Case 1 (p. 24) and some other cases of collective apparitions. One possible explanation, but not the only possible one, could be that the booking clerk at the Edinburgh Festival Office was telepathically aware (by means of the kinds of mechanisms already discussed) that

Miss S knew Miss W and myself (and vice versa), since in principle, according to my theory of telepathy, SMM(complem)s of all memory traces of any particular SMM(brain) (or replicas of these which have descended by repeated replications) can reach every other SMM(brain) of other people, so that there exists the possibility of universal telepathy. In practice this may not take place, because SMM(brain)s that are tightly bound to ordinary matter brains may only have very restricted responses towards incoming telepathic signals. If the clerk, in the case under discussion had the telepathic knowledge discussed (as a possibility) then he could have issue a ticket to Miss S close to Miss W and myself (although the clerk need not have been conscious of this knowledge, which may only have existed as activated memory traces in his SMM(brain)). It should be stressed that the case involved a symbol coincidence 'acquaintance with Miss S,' since both Miss W and I had known Miss S without knowing of each other's knowledge (at least consciously).

Case 52

On Monday 26th March 1984 I watched a TV programme of the BBC *Horizon* series, entitled 'Signs of Apes and Songs of the Whale.' On Wednesday morning 28th March 1984 I talked about this programme to Mr RMW. We talked about language acquisition and concept acquisition in animals, children and human adults. In the course of the conversation I mentioned Kaspar Hauser as a case of very much delayed language acquisition. I had on that day not yet seen *The Times* (London) which was delivered to my house after my departure in the morning to the University. On returning home that night I looked at *The Times* of 28th March 1984 to see what was on the TV. On Channel 4 a programme was titled 'The Enigma of Kaspar Hauser,' was to be shown. It was a German film account of the story of Kaspar Hauser.

Once more it could be suggested that by clairvoyance I became aware of the programme of Kaspar Hauser, as printed in *The Times* (or elsewhere) and that this could have prompted me to talk about Kaspar Hauser in my conversation with Mr W that same day (since I have very rarely ever thought about Kaspar Hauser).

Case 53

Early in 1984 I mentioned to Mrs. S of Newcastle upon Tyne that Monsieur JB of Geneva, a second cousin of mine, must be about 70 years old. She thought that I was wrong. Having been out of touch with Monsieur B for several years, I decided to find out and wrote to him on 26th June 1984. I received a note from Mrs S, written on 27th June 1984, which enclosed a printed address from Grand Rabbin Alexandre Safran of Geneva, in honour of Monsieur B's 70th birthday. Mrs. S had just received this address from her mother who lived (then) in Geneva and who knew Monsieur B.

In this SYNEX Monsieur B's Birthday is the central theme and symbol. I could have become aware, for instance, of Monsieur B's birthday on 26th June 1984 by telepathy between Monsieur B and myself. Alternatively I could have perceived, by pure clairvoyance, the address from Grand Rabbin Safran while in transit in its envelope, and this could have prompted me to write to Monsieur B.

Case 54

According to C. G. Jung (1955, p. 21) the astronomer Camille Flammarion, in his book *The Unknown* (London, 1900), reports on pp. 194 ff that:

A certain Monsieur Deschampes, when a boy in Orléans, was once given a piece of plum-pudding by a Monsieur de Fortgibu. Ten years later he discovered another plum-pudding in a Paris restaurant, and asked if he could have a piece. It turned out, however, that the plum-pudding was already ordered by M. de Fortgibu. Many years afterwards M. Deschampes was invited to partake of a plum-pudding as a special rarity. While he was eating it he remarked that the only thing lacking was M. de Fortgibu. At that moment the door opened and an old, old man in the last stages of disorientation walked in: M. de Fortgibu, who had got hold of the wrong address and burst in on the party by mistake.

Here the coincident symbols are M. Deschampes, M. de Fortgibu and plum-pudding. One can again explain these temporal coincidences in terms of psi-mechanisms. Pure clairvoyance could have been instrumental in directing in each case M. de Fortgibu and M. Deschampes to the locality of the plum-pudding, which, apparently, they both liked and which, at that time, seems to have been a rarity in France. Finally, telepathy between M. Deschampes and N. de Fortgibu could have ensured that they turned up at the potential plum-pudding-providing sources at the same time.

Case 55

The following may be a case of pure telepathy rather than a SYNEX. In August 1984 I attended the International Conference on Systems Research, Informatics and Cybernetics at Baden-Baden, Germany. At the conference was also Professor Joachim Lambek, a very well known Canadian algebraist, and a professor at McGill University, who was not really participating in the conference, but had only come to see somebody. I had never met Lambek before, and sat opposite him on Sunday for dinner. I told Lambek that I had been interned by His Majesty's Government during the war and was shipped to Canada (where I was behind barbed wire at Sherbrooke in 1940). He told me that he had also been interned at Sherbrooke. I asked him whether, like myself, he had been a Jewish refugee. He replied that he was, and told me he came from Leipzig, which was also my birth place and place of residence for seventeen years. He had lived in the Funkenburg Strasse, very close to where I had lived in the Lessing Strasse.

Here we have a case with multiple symbol coincidences, namely Jewish refugees, Internment in Sherbrooke camp, residence in Leipzig. Since hundreds of Jewish refugees were interned at Sherbrooke, the coincidence is not great. Yet, the fact that I mentioned my internment in Sherbrooke to Lambek, that he told me he came from Leipzig, could have been brought about by telepathic prompting, i.e. telepathic interaction (at a partly unconscious level) between Lambek's and my SMM(brain)s, according to the present model. It could be that because of telepathic interaction (of an unconscious kind) I joined Lambek for dinner, because my SMM(brain) could already have registered the relevant symbol coincidences, the symbols being stored as memory traces.

Case 56

The following case is an apparent SYNEX, taken from Myers (1903, vol.1 pp. 660-1) and is quoted by him from the book *Over the Teacup* by Oliver Wendell Holmes (3rd ed. 1891, p. 12). Myers states that:

We are told in the introduction that a part of the book containing the cases was written in March 1888.

Holmes wrote:

I relate a singular coincidence which very lately occurred in my experience...I will first copy the memorandum made at the time:

'remarkable coincidence. On Monday April 18th, being at table from 6.30pm to 7.30 with - and - [the two ladies of my household], I told them of the case of 'trial by battle' offered by Abraham Thornton in 1817. I mentioned his throwing his glove, which was not taken up by the brother of the victim, and so he had been let off, for the old law was still in force. I mentioned that Abraham Thornton was said to have come to this country, and I added 'he may be living near us for aught I know.' I rose from table and found an English letter waiting for me, left while I sat at dinner. I copy the first portion of this letter:

20 Alfred Place West (near Museum),

South Kensington, London, S.W. April 7th 1887

'Dr O.W. Holmes, - Dear Sir, - In travelling the other day I met with a reprint of the very interesting case of Thornton for murder 1817. The prisoner pleaded successfully the old Wager of Battle. I thought you would like to read the account, and send it with this. -

Yours faithfully Fred Rathbone'

(Mr Rathbone was a well known dealer in old Wedgwood and eighteenth century art, whom Mr Holmes had met in England, but from whom he was not expecting any communication. Mr Holmes was a well known lawyer.)

This case could be interpreted in terms of recognition of the contents of the letter by Mr Holmes by pure clairvoyance, and this recognition (although unconscious) could have provided the stimulus to Mr Holmes' SMM(brain) (and from there via his ordinary matter brain) for telling the two ladies about the Trial by Battle. I noted already in **Cases 2, 39, 42, 45, 47, 50** and **52** that apparently some people can perceive by means of pure clairvoyance the contents of shut newspapers, sealed letters, closed books (etc.) and the present Case 56 is a further good example suggesting this apparent ability.

4.2 Some conclusions

I conclude that all of the numerous SYNEXs that I have cited and discussed in detail can, in principle and in practice, be explained in terms of psi-phenomena and their postulated mechanisms. I am not claiming that the explanations that I gave of SYNEXs in terms of psi-phenomena are the only possible ones, and in some cases, at least, more or less obvious, alternative explanations in terms of psi-phenomena are valid. Also, occasionally certain coincidences of symbols could be pure chance, although psi-explanations may seem more convincing in many of the best cases (e.g. **Case 45**), if, as I have tried to show, psi-phenomena could have physical explanations in terms of a physics of Shadow Matter which could be linked to superstring theory in theoretical physics. With all respect to Jung one does not have to invoke an acausal connecting principle which does not belong to the realm of science (whereas Shadow Matter potentially does). Jung's views on the role of acausality were influenced by Schopenhauer's views on closely related topics. He, in turn, was supposedly influenced by Kant's essay 'Dreams of a Spirit Seer' (see Jung, 1955, p. 19 footnote 20). I believe that Jung was mistaken when he tried to interpret psi-phenomena in terms of an acausal connecting principle (see Jung, 1955, pp. 22-3) instead of recognizing that psi-phenomena might, ultimately, be explicable in terms of principles of theoretical physics, as is suggested in this book.

Notes to chapter 4

1. Although I remembered that the invitation was to the Palazzo Fonte de Napoleon in Elba, I did not know that this was near Poggio, and Miss M. Worswick, years later, helped me to find this out.

2. I have suggested elsewhere (Wassermann, 1982b, 1982c) how important evolutionary adaptations could have been brought about non-randomly by purely physical and biophysical processes, so that one can, apparently, avoid the anti-mechanistic acausal factor of Hardy *et al* (1973).

5
The
Possibility
of Survival
of the Human Personality
For At Least Some Time
After Bodily Death

5.1 Materialism and the Possibility of Survival

If there should indeed be survival of the human personality for some time (or possibly permanently) after bodily death, then it is a horrific thought to envisage that even pathological mass murderers, like Adolf Hitler, could be encountered in a 'hereafter'. Yet, the out of the body experiences (OBEs) often associated with near death experiences (NDEs), and already discussed in chapter 1, as well as 'death bed experiences' studied by Osis and Haraldsson (1977), suggest that some people who had already a supposed foretaste of quasi-survival seemed to enjoy it so much that they did not want to return to their ordinary earthly existence. If so, then brutes, like Hitler and his criminal followers (like Eichmann and Himmler), far from ridding the universe, as they thought, of many millions of Jews (and hundreds of thousands of gipsies), whom they murdered, may have helped to transport these Jews, etc., from their miserable existence in German-occupied Europe into a world where the most insignificant Jew from a Polish ghetto counts as much as the once mighty Führer who dominated much of Europe. Of course, I am not writing this to encourage future tyrants to slaughter vast masses of people, because this might transform them into environ-mental states where they are better off than living on Earth. Certainly tyrants must be fought tooth and nail, whether they are called Hitler, Stalin or Saddam Hussein.

It is often thought that survival of the human personality after death of the ordinary matter body is incompatible with materialism. This, however, is not the case, although Osis and Haraldsson[1], and countless philosophical interaction dualists, who believe in the ordinary matter human body being inhabited by a non-material soul, seem to think so. Thus, Osis and Haraldsson (1977, p. 9) assert correctly that:

According to the scientific world view, the universe is exclusively material reality,

144

built only of matter and energy known to physics. Mind and consciousness are mere by-products (epiphenomena) of the physical organism and therefore cannot exist without it. This is still the dominant viewpoint of the scientific community.

Indeed, it is also the point of view of most of the medical fraternity, who, with some notable exceptions, believe that after death you cease to exist. This belief, as we have seen, originates from the view of some, perhaps many, physicists, and many medical people, that physics has now completely explored the world of matter and is familiar with all knowable material entities. This view, however, at most, if at all, applies only to the world of ordinary matter and not to the physically feasible world of Shadow Matter (see section 1.1), and not to a world which (like ours) possibly is composed of both ordinary matter and Shadow Matter, as already discussed extensively in earlier chapters. I noted already that consciousness could be an epiphenomenon (i.e. by-product) not of ordinary brain matter, but of changes of states of Shadow Matter material constituents of the SMM(brain) which, by hypothesis, is normally, during life, bound to the ordinary matter brain. The hypothesized SMM(brain)s of people are also material structures and could survive the disintegration of ordinary matter brains after death. The disintegration of the ordinary matter brain after death does not logically entail disintegration of the SMM(brain) presumably, previously bound to that ordinary matter brain, any more than nearly intact removal of a stamp from an envelope and destruction of that envelope involves logically the destruction of that stamp.

Just as in the world of ordinary matter there exist some structures of immense stability (e.g. fossils), while other structures (such as human ordinary matter brains) decay after death following a relatively short life-span, rarely exceeding 90 years, so something analogous could apply to the world of Shadow Matter structures. For instance, in the world of ordinary matter we have ancient Greek temples and century-old cathedrals which have survived as viable structures, subject to some repair and man-made preservation measures. Although in the world of Shadow Matter we could have structures such as SMM(complem)s with a high turnover, and with a short life-time per SMM(complem) copy, it could be that, just as in the world of ordinary matter, there exist also structures made entirely out of Shadow Matter that are very stable and could exist for years, decades, and possibly indefinitely. Such structures could be stabilized by specific Shadow Matter stabilizer substances and SMM(body)s and their SMM(brain)s could be examples of such stable structures in the Shadow Matter world. Thus, whereas ordinary matter brains decay after death (and likewise other parts of ordinary matter bodies, except skeletons), unless artificially preserved by stabilizers (as in anatomy schools), SMM(brain)s could be stable structures which are stabilized by appropriate, naturally occurring, substances in the Shadow Matter world (see also p. 30 of section 1.3).

Hence, as a metaphysical creed (which it is) mechanistic materialism could hypothesise that the universe consists only of matter, but that this matter comprised ordinary matter and Shadow Matter which can be studied by theoretical physicists, and that consciousness is an epiphenomenon of Shadow Matter and not, as hitherto believed, of ordinary matter. In this way the possibility of survival of a conscious SMM(brain) after death of the ordinary matter body to which it was previously bound, (together with the rest of the Shadow Matter body, i.e. SMM(body) detached from the ordinary matter body), is a possibility fully compatible with mechanistic materialism.

It might be asked: in which way does a SMM(brain) differ from the notion of a 'soul' invoked by religious people of the most varied kinds? The answer is that SMM(brain)s, like the Shadow Matter of which they are composed (by hypothesis) belong to the world of science (see section 1.1) and have scientific properties, e.g. Shadow Matter can form bound states with ordinary matter (see the Introduction), whereas the soul, as postulated by philosophical idealists, has no known or postulated scientific properties. On the other hand ancient materialist philosophers, like Epicurus, suggested that the soul could consist of matter of a special kind (see Long 1974; Rist, 1972), so that the interpretation of the SMM(brain) as a quasi-soul would be a redundant, but interesting exercise in neo-epicurian materialism.

5.2 OBEs and death-bed experiences as putative evidence for the survival of the human personality after death and apparitions of dead people

I believe that OBEs provide good prima facie evidence that there could exist a SMM(body) (which includes an SMM(brain)) which is detachable from the ordinary matter of the human body (see Introduction pp. 13-4). However in an OBE during the life-time of the known ordinary matter body the essentially detached SMM(body) remains residually attached by a reported 'cord,' to the ordinary matter body, as mentioned earlier, the cord is presumably made of Shadow Matter. It is not a far step from OBEs, based (by hypothesis) on essentially detached SMM(body)s, to assume that when death occurs the SMM(body) may become completely detached from the dead ordinary matter body, and survive either for a limited time, or possibly indefinitely with the help of Shadow Matter stabilizer substances (see section 5.1).

Before I proceed I shall cite another case, similar to **Case 17**, in which an OBE is combined with travelling clairvoyance and a reciprocal apparition. This may help the reader to appreciate that such complicated cases, involving OBEs, jointly with other psi-phenomena, are not isolated.

Case 57

This case, like Case 17, was reported by Hart (1956, p. 176) in his collection of interesting case histories. It shows that although the phenomena discussed are rare, they have repeatable class characteristics.

Some time before 1907, a well-known physician of New York City (who was known personally to I. K. Funk, the editor and publisher) was on a river steamer travelling from Jacksonville to Palatka, Florida. He had been having some curious sensations of numbness and of psychological detachment for some days. During the night on the steamer he found that his feet and legs were becoming cold and sensationless. He then 'seemed to be walking in the air' with intense sensations of exhilaration, freedom and clarity of mental vision.[2] In this state he thought of a friend more than a thousand miles distant. Within minutes he was conscious of standing in a room where the gas jets were turned up, and the friend was standing with his back towards him. The friend turned suddenly, saw him and said 'What in the world are you doing here, I thought you were in Florida?' and he started to come towards the appearer. The appearer heard these words distinctly but was unable to answer.

He then had an ecstatic experience of a life beyond the consciousness of time and space. But he decided to return to Earth. He saw his body, propped up in bed as he had left it, but retained the consciousness of another body in which matter of any kind offered no resistance. Then he re-entered his physical body.

In the OBE part of this case there is the felt 'consciousness of another body to which matter of any [ordinary] kind offers no resistance.' This quasi-material body, subsequently identified with the hypothesized SMM(body), was already reported early in this book (see pp. 29, 31, & 37), and in the OBE of **Case 5** (p. 45 ff), Dr Wiltse reported also (see p. 47) that:

> As I turned, my left elbow came into contact with the arm of one of the two gentlemen who were standing in the door. To my surprise, his arm passed through mine without apparent resistance, the severed parts closing again without pain, as air reunites.

So here and in **Case 57** we have not only repeated class characteristics of OBEs, namely the reported quasi-material body, but we have the further repeated class characteristics that this quasi-material can pass through ordinary matter bodies and vice versa. The remainder of Case 57 can be explained as in Cases 15 and 17. The physician in Case 57 became aware also of a friend 'more than a thousand miles distant,' just as Mrs Wilmot (in **Case 15**) became aware of her husband more than a thousand miles distant; and just as the friend became aware of the physician, so Mr Wilmot (and his friend Mr Tait) became aware of Mrs Wilmot. So we are dealing here with perfect similarities. I explained already, in conjunction with Case 15, how such reciprocal apparitions (i.e. paranormal perceptions) could be produced in terms of the psi-mechanisms of my theory.

If my theoretical model is valid, then, whereas there could be a continual turnover of SMM(complem)s, even of SMM(complem)s which in complexity and performance capacities resemble SMM(brain)s (see my interpretation of Case 15) and SMM(body)s, by contrast SMM(brain)s could be very stable (and more generally SMM(body)s) and exhibit no turnover. This would be consistent with the hypothesis that SMM(brain)s, unlike ordinary matter brains, do not experience any turnover during the life-time of an individual (see p. 30 (section 1.3)). This, in fact, would be a requirement if, as my theory suggests, contrary to Crick (1984)), SMM(brain)s, and not ordinary matter brains, are the seats of very stable memory traces. There is, however, no logical necessity for a SMM(brain)'s permanence in order to act as a memory trace repository. For all we know, there could be structures analogous to molecules in the Shadow Matter world, and these structures could, in some cases, act like the memory molecules postulated by Crick (1984) for the ordinary matter world. It must be stressed that in this book I have not exclusively postulated Shadow Matter in order to explain memory traces, but rather in order to explain a host of diverse psi-phenomena and possibly also ordinary psychological phenomena. Accordingly, memory-representing structures of a SMM(brain) are not introduced on an ad hoc basis in order to explain just everyday memories. They are also introduced in order to explain the formation of memory traces during OBEs. In addition, my theory implies that SMM(complem)s which are complementary to SMM(brain)s can also form memory traces. Otherwise it would be difficult to explain within my theory cases such as 15, 17 and 57. Thus, SMM(complem)s of SMM(brains)s, which have a high rate of turnover, could also form, and pass to their descendants, memory traces. By contrast, SMM(brain)s which, because of their association with a personal identity, are assumed to be single-copy structures, do not replicate (and the same applies, more generally, to SMM(body)s and SMM(object)s).

I must now deal with the 'super-ESP' hypothesis (see Gauld, 1977, p. 616 for related remarks). According to this hypothesis a combination of far-reaching pure clairvoyance with telepathy could, allegedly, always explain evidence pointing to 'survival' of the human personality after death. Yet, those who have considered the super-ESP hypothesis did not offer a physical theory for explaining clairvoyance, and telepathy. By contrast, the present theoretical model explains both clairvoyance and telepathy in similar terms, namely in terms of SMM(object)s, SMM(brain)s and SMM(complem)s and their dynamics. Moreover, as noted, within this theory it remains possible that SMM(brain)s, and with them conscious human personalities, could survive ordinary matter brain death. Hence, here we have a fully mechanistic materialistic theory which permits us to explain all kinds of aspects of clairvoyance, telepathy, 'out of the body experiences,' synchronicity experiences, and so forth, to an extent that could provide alternatives to the survival hypothesis in most, or all, cases of alleged survival. Nevertheless, the same theory could also make survival plausible in terms of the same kind of mechanistic materialism. The present mechanistic explanations of clairvoyance combined with telepathy go beyond the expositions of Gauld (1977), Thouless (1984) and Gardner Murphy (1945a, 1945b and 1945c) who each surveyed the evidence bearing on the survival hypothesis, but offered no coherent physical theory for relating survival data to clairvoyance and telepathy (or both) and explaining the latter, in turn, in terms of physics, as is done here.

Case 58

Osis and Haraldsson (1977, p. 13) cite some apparitions of the dead, by referring to Hart (1956). But let me stress, at once, that the human brain, when, by hypothesis, combined with the SMM(brain) has the capacity to synthesize apparitions of the dead, without this implying that the dead are communicating. This was clearly demonstrated by the hypnotically induced apparition of a dead person in Gindes' experiment cited in **Case 3a**. Let us then consider the following case from Hart (1956, p. 158):

In September 1857, Captain Wheatcroft left for India to join his regiment, leaving his wife behind in Cambridge, England. Towards the morning of 15 November she dreamed that she saw her husband ill and anxious. She immediately awoke with her mind much excited. It was bright moonlight, and as she opened her eyes she saw her husband standing beside her bed. He was dressed in uniform, his hands were pressed against his breast, his hair was in disorder, and his face pale. His great black eyes looked at her fixedly and his mouth was contracted. She saw him, in all particulars of his clothing, as distinctly as she had ever seen him during her whole life. He seemed to lean forward with an air of suffering, and he made an effort to speak, but did not utter a sound. This apparition lasted about a minute, then it vanished.

Next morning she told this to her mother and expressed the belief that her husband was either killed or dangerously wounded. Weeks later she received evidence that her husband had been killed at Lucknow on the afternoon of 14 November, about 18 hours before she saw this apparition.

One possibility is that Captain Wheatcroft's SMM(brain) survived, and communicated telepathically with the SMM(brain) of his wife, whose SMM(brain) then synthesized the apparition, much as Gindes' subject in (**Case 3a**, p. 42) synthesized the apparition of her dead brother. Alternatively, Mrs. Wheatcroft's SMM(brain) could, by means of clairvoyance, have become aware of her husband's death on 14th November, although her SMM(brain) only synthesized the apparition of her husband a day later. In either explanation it is assumed that Mrs Wheatcroft's SMM(brain) synthesized the apparition. So in this case there is no strong case for the survival hypothesis over the clairvoyance hypothesis.

Case 59

The following case history is also cited by Hart (1956, p. 160).

John E. Husbands wrote a letter dated 15 September 1886, stating that while sleeping in a hotel in Madeira early in 1885 [later ascertained to be after February 1885] on a bright moonlit night, with the windows open and the blinds up, he felt someone was in his room. On opening his eyes he saw a young fellow about twenty-five dressed in flannels standing at the side of his bed and pointing the first finger of his right hand to the place in which Husbands was lying. Husbands lay still for some seconds, then sat up and looked at the man. He saw his features so plainly that he recognized them in a photograph which was shown him some days later. Husbands asked the apparition what he wanted; the apparition did not speak, but his eyes and hands seemed to tell Husbands that he was occupying the apparition's place. As the apparition did not answer, Husbands struck out with his fist, but did not reach him. As Husbands was going to spring out of bed, the apparition slowly vanished through the door, which was shut, keeping his eyes upon Husbands all the time. On 8 October 1886, K. Falkner wrote that the figure Mr Husbands saw was that of a young fellow who had died unexpectedly some months previously [29 January 1884] in the room which Mr Husbands was occupying. Husbands had never heard of the man or of his death.

One possible explanation would be that the young fellow's SMM(body) including his SMM(brain) survived the death of his ordinary matter body. If so, then the SMM(body) with its SMM(clothes) could have become trapped by (i.e. bonded to) a wall or some object of the room and the surviving trapped SMM(body) could then have emitted replicating SMM(complem)s one of which could be perceived (via sphotons) with the help of Mr Husbands' Shadow Matter eyes, so that Husbands could have perceived the figure represented by the SMM(complem) by means of his SMM(brain). The latter could then have synthesized the apparition by the same machinery that synthesized the apparition in Gindes' case (**Case 3a**, p. 42). Alternatively, it could be assumed, in keeping with my theory, that the young man's SMM(body) did not

149

survive, but that, while he was dying, his SMM(body) emitted many replicating SMM(complem)s, many of which got captured by the walls of the hotel room (or by other objects in the room), where they could subsequently replicate, by the mechanisms described earlier. One of these replicating SMM(complem)s could then, after its emission from the parent template, have interacted with Mr. Husband's Shadow Matter eyes (via spotons), leading then to the same consequences as described for the other alternative given. Since the young man was 'seen' by Husbands in his clothes, it may be assumed that, according to my theory, the SMM(complem)s formed were of the SMM(body) of the young fellow together with the Shadow Matter representation of his clothes. Or, alternatively these clothes could have been hallucinated by Husbands as in Dr Wiltse's 'out of the body experience' (see Case 5, p. 48, where Dr Wiltse suddenly noticed himself clothed, whereas before, when emerging from his ordinary matter body, he seemed to have a 'naked' quasi-material body). More generally, all kinds of hauntings of buildings by people could be explained in similar terms. (See also my general discussion (which did not assume survival of a SMM(brain) or SMM(body)) in section 2.5 and **Case 13** p. 91).

Case 60

The following is an account of a 'haunting' apparition, witnessed at different times by several people, and was presented by Hart (1956, p. 161).

The narrator stated that one July morning in the year 1873, as she opened her eyes from sleep, about 3am, she saw a figure of a woman, stooping down and apparently looking at her. The woman's head and shoulders were wrapped in a common grey woollen shawl. Her arms were folded, and they were also wrapped, as if for warmth, in a shawl. The percipient looked at the woman in horror and was afraid to cry out. After a time whose duration the percipient could not judge, the apparition raised herself and went backward towards the window, stood at the table, and gradually vanished - i.e. grew by degrees transparent. The reporter stated that she was ready to take an oath that she did not mention the circumstances to either her brother or servant.

Exactly a fortnight later, the percipient noticed that her brother was out of sorts at breakfast time. The brother stated that early in the morning he had seen, as distinctly as he saw his sister, a villainous looking old hag, with her head and arms wrapped up in a cloak stooping over him. The first percipient then said that she had seen the same thing a fortnight previously.

About four years later, a boy of four or five years of age, left alone in the drawing room of the house, came out pale and trembling, and said to the narrator's sister, 'who is that old woman that went upstairs?' The sister tried to convince him that there was no old woman, and though they searched every room in the house the child still maintained that the old woman did go upstairs.

As in the previous case one can assume that the SMM(body) (and clothes) of the old woman emitted SMM(complem)s, while the woman was in the house, and that these SMM(complem)s, and/or their descendent replicas, got trapped in the walls or other parts of the house and subsequently acted as 'breeders' of other SMM(complem)s of the same kind (by replications of the kind described earlier). These additional SMM(complem)s were then emitted by the house (e.g. by the walls or furniture) and one of them, or the other, could then have acted as a trigger, when encountering people, for forming apparitions by means of these people's SMM(brain)s, provided these percipients were receptive (see **Case 59**). These percipients' SMM(brain)s could have synthesized the apparition, so as to fit into the surroundings of the house as seen from their point of view. This 'adaptation' or 'fitting into surroundings' occurred also in the hypnotically induced post-hypnotically occurring apparition, reported by Gindes and discussed repeatedly (see **Case 3a**, p. 42). Hence, in this case again, one does not have to assume a 'surviving entity' in the guise of surviving SMM(body) and its SMM(brain), although survival of the old woman's SMM(body) cannot be ruled out as a possible prior cause. One may assume that trapped SMM(complem)s bound to parts of buildings (or to other ordinary matter within or near these buildings) could 'survive' for long periods (or else could replicate and also show turnover) (e.g. to persist at least four years in the preceding case). If, however, SMM(complem)s could 'survive' i.e. stably persist in a form bound to ordinary matter for long periods of time (as distinct from the turnover of free SMM(complem)s postulated on p. 80), then one cannot exclude the possibility that SMM(body)s and their SMM(brain)s could also survive ordinary bodily death, perhaps in a form bound to ordinary matter.

Case 61

The following case involves a multiple simultaneous apparition of a dead person, and may have involved telepathy. It is from the *Journal of the Society for Psychical Research* (London) vol.6, p. 230, and is also summarized by Tyrrell (1953, p. 137) whose summary I shall give:

Two ladies were looking over a church in which was placed the tomb of an old friend of one of their families. This friend had left a sum of money to have glass put into a window in his memory, but his heir neglected his wish. On hearing this story, the narrator of the case says that she felt quite angry and said to her companion; 'If I was Dr. - I should come back and throw stones at it.' 'Just then' she continues, 'I saw an old gentleman behind us, but thinking he was looking over the church took no notice. But my friend got very white and said, 'Come away, there is Dr - !' Not being a believer in apparitions, I simply for the moment thought she was crazy... But when I moved, still looking at him, and the figure before my very eyes vanished, I had to give in.'

As one of the two ladies had known Dr - she must, according to my theory, have formed memory traces of Dr - in her SMM(brain). The mention of Dr -, or the thought about him, could then have induced a hallucination (as in Gindes' case, **Case 3a**, p. 42) of Dr -, and by telepathy a corresponding hallucination, 'fitting equally well into the surroundings' (see my remarks in **Case 60**) could then have been induced in the other lady's SMM(brain) and elaborated by that SMM(brain). Alternatively, the surviving SMM(brain) of Dr -which may have been bound to ordinary matter of his tomb in the church, could have emitted SMM(complem)s, which acted in this case on the SMM(brain)s of one or both ladies. Hence in this case as in the other cases cited above, there exist no compelling reasons that SMM(body)s and their SMM(brain)s must have survived, although that possibility cannot be excluded.

Case 62

The following apparition, coincident with the death of the subject of the apparition, was recorded by Sidgwick (1923) and summarized by Murphy (1945a) as follows [for a fuller account see also West (1954) pp. 25 ff]:

The percipient was Lieut. J. J. Larkin of the RAF, and the apparition was that of one of Lieut. Larkin's fellow officers, Lieut. David M'Connell, killed in an airplane crash on December 7, 1919. Lieut. Larkin reported that he spent the afternoon of December 7th in his room at the barracks. He sat in front of the fire reading and writing, and was wide awake all the time. At about 3pm he heard someone walking up the passage. 'The door opened with the usual noise and clatter which David always made; I heard his "Hello boy!" and turned half round in my chair and saw him standing in the doorway, half in, half out of the room, holding the door knob in his hand. He was dressed in his full flying clothes but wearing his naval cap, there being nothing unusual in his appearance. . . In reply to his "Hello boy!" I remarked "Hello! back already?" He replied, "Yes got there all right, had a good trip. . ." I was looking at him the whole time he was speaking. He said, "Well, cheerio!" closed the door noisily and went out.' Shortly after this a friend dropped in to see Lieut. Larkin, and Larkin told him that he had just seen and talked with Lieut. M'Connell. (This friend sent a corroborative statement to the Society for Psychical Research (London)). Later on that day it was learned that Lieut. M'Connell had been instantly killed in a flying accident which occurred about 3.25pm. Mistaken identity seems to be ruled out, since the light was very good in the room where the apparition appeared. Moreover, there was no other man in the barracks at the time who in any way resembled Lieut. M'Connell. It was also found that M'Connell was wearing his naval cap when he was killed - apparently an unusual circumstance. Agent and percipient had been 'very good friends though not intimate friends in the true sense of the word.'

It could be assumed that M'Connell's SMM(body) and his SMM(brain) survived and generated SMM(complem)s which then moved off and replicated. One of these SMM(complem)s could then have reached Larkin's SMM(brain) and induced the hallucination, which was then fully produced by Larkin's SMM(brain). Alternatively Lieut. Larkin could have become aware clairvoyantly of the accident (though not at a conscious level), via his SMM(brain) and this could have triggered the hallucination-forming machinery of the type referred to in the preceding cases. As in **Case 60**, there occurred, in the present case, a remarkable behavioural 'fitting in' of the apparition into the surroundings and in its response to the percipient (including an auditory hallucination). Since, as once more repeated, back reference to Gindes' case (**Case 3a**, p. 42) indicates that human beings (i.e. their assumed SMM(brain)s) are capable of synthesizing apparitions that fit visually, auditorily and behaviourally very naturally into their environment and that of the percipient, this suggests that this human apparition-synthesizing machinery may have been at work, either with the help of the surviving SMM(brain) or via incoming clairvoyant stimuli. The case certainly presents good evidence for powerful psi-phenomena, since it would be too much of a coincidence that the apparition should be constructed by Larkin's (say) SMM(brain) just at the time of M'Connell's death, and without either clairvoyant psi-communication initiated by Larkin or, perhaps more likely, psi-communication initiated by M'Connell's surviving SMM(brain) and propagated via SMM(complem)s.[3]

Case 63

The following description of an apparition of a dead person comes from Myers (1903, vol.2, p. 371) and the original account came from Mrs Clark, 8 South View, Forest Hall, Newcastle upon Tyne on 6 January 1885. Murphy (1945a, p. 4) condensed the account as follows:

Mrs Clark stated that a young gentleman, Mr Akhurst, had been much attached to her and had wanted to marry her. She became engaged to Mr Clark, however, and later married him. After she had been married to Mr Clark for about two years, Mr Akhurst came to visit them in their home in Newcastle upon Tyne. It appeared that at this time he was still interested in her. Mr Akhurst then went to Yorkshire and Mrs Clark never heard from him again. Three months passed and her baby was born. At the end of September 1880, very early one morning, as she was feeding her baby, 'I felt a cold waft of air through the room and a feeling as though something touched my shoulder. . . Raising my eyes to the door (which faced me), I saw Akhurst standing in his shirt and trousers looking at me, when he seemed to pass through the door. In the morning I mentioned it to my husband.' Mr Clark wrote in corroboration, 'Shortly after my wife had been confined of my second daughter, about the end of September, 1880, my wife one morning informed me she had seen Akhurst about one o'clock that morning. I of course told her it was nonsense, but she persisted and said he appeared to her with only trousers and shirt on. . . ' Upon inquiry, it was learned that Mr Akhurst had died (as a result of an overdose of chloral) on July 12, 1880. A friend said that Akhurst was found dressed in only a shirt and trousers. The interval between death and apparition is thus seen to be about ten weeks.

One feasible explanation of this case is that Mrs Clark's SMM(brain) perceived the dying or dead Akhurst clairvoyantly, and that after a delay of ten weeks this SMM(brain) constructed the apparition. In the repeatedly referred to case of Gindes (Case 3a, p. 42), there also occurred a delay of several months between the hypnotic suggestion and the occurrence of the apparition of the deceased brother of the percipient. One cannot rule out that Akhurst's surviving SMM(brain) and SMM(body), via emitted SMM(complem)s, produced the stimulus for the apparition, but this seems, perhaps, a less economical hypothesis, although I am not at all sure that it is less economical.

Case 64

The following case comes from Myers (1903, vol.2, pp. 27-30). He stated amongst other things that the events related occurred in 1876 and were experienced by a commercial traveller while on a western trip in the USA. He stated:

I had 'drummed' the city of St. Joseph, Mo and had gone to my room at the Pacific House to send my orders, which were unusually large ones, so that I was in a very happy frame of mind indeed. My thoughts, of course, were about these orders, knowing how pleased my house would be at my success. I had not been thinking of my late sister, or, in any manner reflecting on the past. The hour was high noon, and the sun was shining cheerfully into my room. While busily smoking a cigar and writing out my orders, I suddenly became conscious that some one was sitting on my left, with one arm resting on the table. Quick as a flag I turned and distinctly saw the form of my dead sister, and for a brief second or so looked her squarely in the face; and so sure was I that it was she, that I sprang forward in delight calling her by name, and as I did so, the apparition instantly vanished. Naturally I was startled and dumbfounded almost doubting my senses; but the cigar in my mouth and pen in my hand, with the ink still moist on my letter, satisfied myself I had not been dreaming and was wide awake. I was near enough to touch her, had it been a physical possibility, and noted her features, expression, and details of dress etc. She appeared as if alive. Her eyes looked kindly and perfectly natural into mine. Her skin was so life-like that I could see the glow of moisture on its surface, and, on the whole, there was no change in her appearance, otherwise than when alive.

Now comes the most remarkable confirmation of my statement, which cannot be doubted by those who know what I state actually occurred. This visitation . . . so impressed me that I took the next train home, and in the presence of my parents and others I related what had occurred. My father, a man of rare good sense and very practical was inclined to ridicule me, as he saw how earnestly I believed what I stated; but he too was amazed when later on I told them of a bright red line or scratch on the right-hand side of my sister's face, which I distinctly had seen. When I mentioned this my mother rose trembling to her feet and nearly fainted away, and. . . with tears streaming down her face, she exclaimed that I had indeed seen my sister, as no living mortal but herself was aware of that scratch, which she had accidentally made while doing a little act of kindness after my sister's death. She said that she well remembered how pained she was to think that she should have, unintentionally, marred the features of her dead daughter, and that unknown to all, how she had carefully obliterated all traces of the slight scratch with the aid of powder etc., and that she had never mentioned it to a human being from that day to this. In proof, neither my father nor any of our family had detected it, and positively were unaware of the incident, yet I saw the scratch as bright as if just made.

One possible interpretation of this case is as follows. The commercial traveller's mother had, obviously, formed memory traces of the daughter with a bright red scratch on her face, otherwise she could not have recalled the incident. Hence, according to my theory, the mother's SMM(brain) contained memory traces of the daughter with the scratch on her face. Via SMM(complem)s information about these memory traces could have been telepathically transferred from the mother's SMM(brain) to the traveller's SMM(brain), and the latter could then again as in Gindes' case (Case 3a, p. 42) have generated an apparition of the sister with the scratch on her face. Alternatively, one cannot rule out possible survival of the sister's SMM(body) (including her SMM(brain)) as the triggering cause of the apparition, which even in this case was constructed by the traveller's SMM(brain). Yet it is not clear how the sister's SMM(body) could have incorporated the equivalent of the scratch *after* death.

I prefer the telepathic interpretation.

Case 65

The following case of James L. Chaffin (*Proceeding of the Society for Psychical Research* (London), 1926-28 vol.36, pp. 517-24) does not seem as unambiguous as regards survival as Murphy (1945a, pp. 7-8) believed. He wrote that:

James L. Chaffin appeared to one of his sons. Information was conveyed as to the whereabouts of a second will benefiting the percipient. The existence of this second will was not known to any living person. It was found, however, and was accepted as valid in the State of North Carolina, where the Chaffin family lived. Mr Chaffin had been dead for about four years when his son's series of dreams began.

It could be argued that we have here another example (as in **Cases 2, 39-40, 42-4** etc.) where the percipient became clairvoyantly aware of the contents of a sealed (or otherwise hidden) letter (or document), interior of a book, etc., in this case the stowed away printed or written pages of a will. On the other hand, the fact that the information became known first to Mr Chaffin's son (whom the will benefitted) and that Mr Chaffin's apparition was involved would also be consistent, perhaps more consistent, with the hypothesis of the surviving SMM(body) (and its SMM(brain)) of Mr Chaffin. (Philosophically inclined readers, particularly those interested in the philosophy of mind, should notice the apparent intentionality of the supposedly 'surviving' Mr Chaffin Senior.) A surviving SMM(brain) of Mr Chaffin could certainly have provided the required intentionality directed towards both the will and towards his son, whereas clairvoyance may not have the requisite inbuilt intentionality,[4] although this is debatable. The problem is that the physicalistic Shadow Matter theory of this book could explain communication between a 'surviving' SMM(brain) and a 'living' SMM(brain) in terms of the transmission (and emission) of SMM(complem)s, while the same theory could explain clairvoyance in terms of a closely related type of machinery. Yet it looks as if **Case 65** is marginally easier to explain in terms of a postulated surviving SMM(brain) of Chaffin Senior.

Case 66

Again, consider the case, cited by Myers (1903, vol.2, p. 37) taken from a report in the *Herald* (Dubuque, Iowa) February 11th 1891.

It will be remembered that on February 2nd Michael Conley, a farmer living near Ionia, Chickasaw County, was found dead in an outhouse at the Jefferson house. He was carried to Coroner Hoffmann's morgue, where, after the inquest, his body was prepared for shipment to his late home. The old clothes which he wore were covered with filth from the place where he was found, and they were thrown outside the morgue on the ground.

His son came from Ionia, and took the corpse home. When he reached there, and one of the daughters was told that her father was dead, she fell into a swoon, in which she remained for several hours. When at last she was brought from the swoon, she said, 'Where are father's old clothes? He just appeared to me dressed in a white shirt, black clothes, and felt slippers, and told me that after leaving home he sewed a large roll of bills inside his grey shirt with a piece of red dress, and the money is still there.' In a short time she fell into another swoon, and when out of it demanded that somebody go to Dubuque and get the clothes. She was deathly sick, and is so yet.

The entire family considered it only a hallucination, but the physician advised them to get the clothes, as it might set her mind at rest. The son telephoned Coroner Hoffmann, asking if the clothes were still in his possession. He looked and found them in the backyard, although he had supposed they were thrown in the vault, as he had intended. He answered that he still had them and on being told that the son would come to get them, they were wrapped up in a bundle. The young man arrived last Monday afternoon, and told Coroner Hoffmann what his sister had said. Mr. Hoffmann admitted that the lady had described the identical burial garb in which her father was clad, even to the slippers, although she never saw him after death, and none of his family had seen more than his face through the coffin lid. Curiosity being fully aroused, they took the grey shirt from the bundle, and within the bosom found a large roll of bills sewed with a piece of red cloth. The young man said his sister had a red dress exactly like it. The son wrapped up the garments and took them home with him yesterday morning, filled with wonder at the supernatural revelation made to his sister, who is at present lingering between life and death.

I believe that there is nothing supernatural about this case, and that it can be explained in terms of the mechanistic materialistic Shadow Matter theory expounded in this book. One possibility is that the SMM(body) (including the SMM(brain)) of Mr Conley survived the death of his ordinary matter body. Since Conley knew that he had sewn the roll of bills with a piece of his daughter's red dress inside his grey shirt, this information could have been stored in the form of memory traces in Conley's SMM(brain), and these memory traces survived with that SMM(brain). The information could then have been conveyed telepathically, by the mechanisms stated earlier, from Conley's surviving SMM(brain) to his daughter's SMM(brain). Likewise, just as in an 'out of the body experience' (OBE) Mr Conley's surviving SMM(body) (including his SMM(brain)) could have left his ordinary matter body (either at death or soon after) and perceived his ordinary matter body, dressed in the burial garb, and his surviving SMM(brain) could then, again by telepathy, have conveyed this information to his daughter's SMM(brain).

Alternatively, Conley's daughter could have perceived clairvoyantly Conley in his death garb, and her SMM(brain) could earlier have registered, unconsciously at that time, Conley's actions in getting hold of part of her red dress and sewing the roll of bills into it, the SMM(brain) acting again by means of clairvoyance. Out of this information the daughter's SMM(brain) could then have constructed the hallucination. In both alternatives the SMM(brain) of Conley's daughter would have to construct the apparition and its message from the available information which was either telepathically (from the surviving SMM(brain) of Mr. Conley) or clairvoyantly conceived by her SMM(brain). Whichever of these alternatives is valid, they provide, quite by the way, overwhelmingly powerful suggestions that psi-phenomena exist and could have operated by the mechanisms suggested earlier in this book.

Even if James Randi (1991) shouts from the roof tops that psi-phenomena do not exist, then we must allow him to live in a fool's paradise. As *The Sunday Times* 5th January 1992, p. 4 notes, Randi 'has made a career of debunking those who claim psychic powers. . . ' I am inclined to think that one can make a lot of money by misleading certain people into thinking that phenomena that patently exist cannot and do not exist. I believe that Randi is not alone in propagating his philosophy, if that is what you wish to call it. Some months ago, I visited an old friend of mine. When I told him that I was writing a book on psi-phenomena, he threw himself back in his chair and exclaimed 'such phenomena are impossible!' I am sure that people like him devour Randi's writings.[5]

Let me now turn to death-bed experiences. Osis and Haraldsson's (1977) book is based on three surveys involving a thousand cases observed by medical staff and nurses. In their chapter 5, titled 'What the Dying See' it seems that many of the death-bed apparitions share some of the features of OBEs, although the most common feature seems to be that the dying person perceives an apparition of a dead relative or dead friend who seems to come in order to fetch him or her.

Case 67

Osis and Haraldsson (1977, p. 38) state that:
In one of our cases, a nineteen-year-old girl saw her dead father coming for her, in addition to seeing bright lights[6] and other people. In spite of the experience of light, a typical quality of otherworldly visitors, the girl was very scared. She called to the nurse, 'Edna hold me tight,' and then died in her arms. Thus, the apparition and the patient seemed to be at cross purposes because she did not want to go. This certainly does not look like any form of self-suggestion or wish fulfilment.

Case 68

Here is a somewhat atypical case from Osis and Haraldsson (1977, p. 44).
A college-educated Indian man in his twenties was recovering from mastoiditis. He was doing very well. He was going to be discharged that day. Suddenly at 5am he shouted 'someone is standing here dressed in white clothes. I will not go with you!' He was dead in ten minutes. Both the patient and the doctor expected a definite recovery.

It would be rash to assume that death-bed apparitions, like the girl's dead father in **Case 67** or the figure dressed in white in **Case 68**, are surviving entities which come to 'fetch' the dying person. They could be simply hallucinations produced by the SMM(brain) of the dying person. If, as my theory assumes, in death, as in an OBE, there occurs a separation between SMM(body) (including the SMM(brain)) and the ordinary matter body of the dying person, then we could assume that as death approaches there occurs a weakening of the (Shadow Matter)-(Ordinary matter) bonds between SMM(brain) and ordinary matter brain, thereby giving greater autonomy to the SMM(brain). In this state then, as in some OBEs, the SMM(brain) could synthesize some hallucinations (see the hallucinated suit of scotch material in the OBE of **Case 5**, p. 47-8) and the death-bed apparitions of dead people may be no more than self-generated hallucinations, manufactured by the still living SMM(brain), rather than evidence of surviving SMM(body)s and their SMM(brain)s. In fact, Osis and Haraldsson (1977, p. 63) noted that in numerous death-bed experiences the apparition was a religious figure (e.g. Christ) of the religion of the dying person. This makes it even more probable that the apparition was synthesized by the SMM(brain) of the dying person, particularly if the dying person was brought up in the religious tradition to which the apparition belonged. I have cited already many case histories showing that human beings could, on the basis of clairvoyance or telepathic cues, or even merely as the result of hypnotic suggestion (pp. 42 ff, **Case 3a**), synthesize immensely complex apparitions. None of which provide compelling evidence for the survival of human personalities after death of the ordinary material body (if these personalities are, say, related to SMM(brain)s). Similar remarks apply to death-bed apparitions.

Significantly, Osis and Haraldsson (1977, p. 64) note that apart from death-bed hallucinations of (mostly close) relatives 'the second largest category was that of religious figures. On the whole Christians tended to hallucinate Angels, Jesus or the Virgin Mary, whereas Hindus would mostly see Yama (the god of death) one of his messengers, Krishna, or some other deity.' The fact that death-bed apparitions also occurred to people who did not expect to die (see **Case 68**) suggested to Osis and Haraldsson (1977, p. 88) that these apparitions could be of extrinsic origin. This, however, does not follow, since, although the percipient did not expect to die, there could, unknown to him or her, already have occurred a weakening of bonds between SMM(brain) and ordinary matter brain, thereby substantially facilitating the synthesis of endogenous hallucinations. This, of course, does not exclude that some of these hallucinations, although (by hypothesis) SMM(brain)-manufactured, could be telepathically or clairvoyantly triggered by extrinsic sources. The fact that, after having had apparitional death-bed experiences, some of the dying subjects may feel very serene, does not imply that the apparitions came from a 'hereafter,' since pacifying effects could be produced by the still 'living' SMM(brain)'s cognitive machinery. That, notwithstanding fearful pains from terminal cancers, or heart attacks, dying patients can terminally become very happy and serene seems typical of numerous death-bed experiences. This could be the result of the much weakened bonding between the ordinary matter brain and the SMM(brain) which may increasingly exclude the transmission of pain-arousing signals (via gravitons) from the ordinary matter brain to the SMM(brain). The great composer Verdi must have had some idea about such last minute states of elation when he composed his *Traviata* (and so had his librettist). In the last act the moribund Violetta Valéry, suffering from terminal tuberculosis, suddenly becomes very elated, just before she collapses and dies.

5.3 Further evidence relevant to the survival hypothesis

In communications of factually accurate statements from mediums, while in a state of trance, it is often alleged that 'messages' pronounced by the medium come from deceased people. Yet, in many, perhaps all, cases these 'messages' could have been synthesized by the SMM(brain) of the medium with the help of, say, telepathic communication between the medium and living people, or with the help of clairvoyance. The living people concerned could be the 'sitter' who receives the 'message' (or somebody far removed). (In some cases the SMM(brain) of the medium clearly seems to rely on clairvoyant information.) For instance, the renowned trance medium Mrs Piper told one 'sitter,' Mr Clarke, (via Mrs Piper's 'control' Dr Phinuit - often thought to be a 'secondary personality' of Mrs Piper) about 'tickets with figures stamped in red' at a time when Mr Clarke, although not actively thinking about the tickets, had repeatedly thought about this topic (Leaf, 1889-90, p. 572). This suggests telepathic communication of information from Mr Clark's memory traces to Mrs Piper's assumed SMM(brain).

Again, the clairvoyant faculties of some mediums (and non-mediums) seem impressive. One such medium was Mrs Osborne Leonard. Radclyffe-Hall and Troubridge (1918-19, pp. 506-21) had sittings with Mrs Leonard, where a communicator called AVB. cited states of affairs unknown to the investigators, where Mrs Leonard could have received the information by clairvoyance. Murphy (1945a, p. 11) summarized the case thus:

Case 69

In one such instance AVB describes the home of a friend she had known well during her lifetime, this home being unknown to the sitters and to the medium. After describing the house, AVB through Feda (Mrs Leonard's 'control') [presumably a secondary personality of Mrs Leonard] refers to some things hanging on the walls of a room 'which things are long in shape, though not all long in shape; they are, however, nothing to do with pictures, and one of them is said to have been *dried*.' Feda then speaks of one or two portfolios containing designs and drawings and of a collection of books pertaining to semi-civilized peoples, and of a 'very old chest.' Through correspondence with the owner of the home, the investigators learn that he had hanging on the walls of his vestibule 'weapons and stuffs from the Sudan and elsewhere, many of them long in shape.' Also a *dried* crocodile from the Nile. He also had a portfolio containing drawings and sketches for the alteration and decoration of the vestibule where the dried crocodile hung. In his library he had a collection of books on Central Africa, the Sudan etc. Finally he had 'a very old chest' - an old Italian Casonne. Further inquiries elicited the fact that AVB in her lifetime had been interested in all the items referred to; she had, for example, seen the sketches for remodelling the vestibule and had discussed them with her friend.

Here, again, we have the possibilities of several alternative explanations in terms of my theory. First there is the possibility that the SMM(brain) of AVB survived, and, by telepathy, communicated to Mrs Leonard's SMM(brain) an account of her information (stored in AVB's SMM(brain) by memory traces), namely information concerning her friend X's house. Alternatively, Mrs Leonard's SMM(brain) could by telepathy have derived all the information referred to from the SMM(brain) of the friend X, who stored the information as memory traces, and the friend X was unknown to Mrs Leonard. One cannot rule out that Mrs Leonard's SMM(brain) could, by 'pure clairvoyance,' have received information about the objects in X's house. That living people may have clairvoyant capacities of the most astonishing kind was suggested, in many cases, in my explanations of numerous SYNEXs in chapter 4, provided one interprets these SYNEXs in terms of psi-phenomena, instead of 'explaining' them as chance coincidences, as debunkers, such as James Randi, would undoubtedly do. Debunkers often believe that the most naive 'explanations' are the most plausible and that all plausible mechanisms must be obvious to the man in the street, and that complex mechanisms which are not obvious to the debunkers, and cannot be understood by them, simply cannot exist. There exist, of course, many cases of apparent 'pure clairvoyance' which do not involve SYNEXs. Thus, in the *Radio Times* (London, 24-30 September 1983, pp. 92 ff) Nigel Lewis wrote about the sensitive Pat Price, a former building contractor and police commissioner, who was a subject employed by Russell Targ for investigation. Price, apparently would 'describe buildings he was "seeing" clairvoyantly in the most amazing and accurate detail.'

Not surprisingly the secret services of the (former) superpowers have tried to utilize such psi-abilities of a few gifted subjects in order to pry into buildings, documents etc., of other nations. Lewis (1983, p. 92) noted that the military establishment take psi seriously 'to the tune, over the last decade of some £3m in the American camp and (according to some accounts) up to £30m in the Soviet.' A corollary of this is that if the former superpowers had taken the arguments and 'demonstrations' of debunkers like James Randi and his ilk seriously, they would not have relied on putatively existing psi-faculties.

Case 70

The following case reveals also likely striking telepathic powers, resembling those of Case 69 (provided one wishes to interpret Cases 69-70 in terms of telepathy rather than survival). The case is cited by Murphy (1945a) and is based on Radclyffe-Hall and Troubridge (1918-19). I shall cite only excerpts from Murphy's summary of the case.

Miss Radclyffe-Hall had a friend, Daisy Armstrong (pseudonym), who had lost her husband during the First World War. Daisy wrote from the Near East to Miss Radclyffe-Hall (MRH) if she would try to obtain for her through Mrs. Leonard some evidence from her husband... At a sitting held on February 21st 1917... after many other references to past events and interests, subsequently verified by either Daisy or her sister, the communicator purporting to be [Daisy's] father says; 'There were two of us that stood in the same relation to Daisy. . . two of us did stand in the same relation to Daisy, with a slight difference. . . Do you follow me?' Feda has referred just before this to a man 'writing in jerks,' and has described at length a machine in this man's room. 'It's nearly all made of some dark-coloured metal. . . a big thing on a stand. . . like a rolly thing or rod something seems to rise up, something that looks curved.' All of this Daisy, in the Near East, recognized as applying perfectly to her adopted father, the Rev Bertrand Wilson whom she believed to be alive, in England. But it is subsequently learned that he had died three days before this sitting had taken place. The adopted father was a composer and sat all day at his table writing music. In the next room he had a printing press. The investigators wrote: 'An inspection of the Excelsior and Model Hand-Printing Machines has revealed that Feda's description of the machine. . . was very near the mark indeed.' Neither of the investigators knew that Daisy had an adopted father (MRH had seen little of her for many years and Lady Troubridge had never met her), nor did they know that such a person as Rev Mr Wilson existed. A perfect stranger to the entire group concerned had to be approached in order to ascertain the date of the adopted father's death.

All the information concerning Daisy's adopted father (Mr Wilson) and the printing machine etc., was known to Daisy at the time of the sitting, and could have been telepathically transmitted from her SMM(brain), via SMM(complem)s, to the SMM(brain) of Mrs Leonard (relying on the SMM(complem) replication mechanism postulated earlier, which permits efficient communication of psi-information over very long distances). There is no need to invoke intervention of the surviving SMM(brain) of Mr Wilson and/or the surviving SMM(brain) of Daisy's dead husband, although this possibility cannot be ruled out. The fact that Daisy was hundreds or more miles away from the sitters, at the time of the sitting, does not exclude telepathy, since psi-communication can take place over thousands of miles (see section 2.2.2 p. 81).

Perhaps we should also pity those warped debunkers of psi-phenomena who invariably declaim that in such cases the trance medium obtained all the relevant information by sensory channels, however impossible this may seem or be. Debunkers more often than not cherish to rely on arguments of the species 'what could happen if pigs could fly,' which are not the same as scientific hypotheses.

I shall now turn to so-called 'Book Tests,' which concern evidence allegedly not known to any living person (Murphy, 1945a, p. 14), unless known to the trance medium concerned by pure clairvoyance. The following case, published by Mrs Sidgwick (1920-21) is included in Murphy's (1945a) collection of cases.

Case 71

Murphy writes:

The sitter was Mrs Hugh Talbot and the purported communicator was her husband. [The trance medium used was Mrs Leonard]. Mrs Talbot reported (in part) as follows:

Suddenly Feda (Mrs Leonard's 'control') began a tiresome description of a book, she said it was leather and dark, and tried to show me the size. Mrs Leonard showed a length of eight to ten inches long with her hands, and four or five inches wide. She (Feda) said 'it is not exactly a book, it is not printed. . . it has writing in. . . There are two books, you will know the one he means by the diagram of languages in front. . . Indo-European, Aryan, Semitic languages. . . A table of Arabian languages, Semitic languages.' It sounded absolute rubbish to me. I had never heard of a diagram of languages and all these Eastern names jumbled together sounded like nothing at all, and she kept on repeating them and saying this is how I was to know the book and kept on and on 'will you look at page twelve or thirteen.' [In the earlier part of the sitting the communicator had repeatedly asked the sitter to believe that life continued after death and that he did not feel changed at all.] Mrs Talbot reported that the next day she found two old notebooks which had belonged to her husband and which she had never cared to open. A shabby black leather one corresponded in size to Feda's description. 'To my utter astonishment, my eyes fell upon the words, "Table of Semitic or Syro-Arabian Languages", and pulling out the leaf, which was a long and folded piece of paper pasted in, I saw on the other side "General table of Aryan and Indo-European languages." On page thirteen of this notebook was an extract from an anonymous work entitled *Post Mortem*. It describes the sensation of a person who realizes that he is dead, and of his meeting with a deceased relative.'[7]

On the face of it this case looks like possible survival evidence consistent with the hypothesis that the SMM(body) (including the SMM(brain)) of Mr. Talbot survived the death and decay (or possible cremation) of his ordinary matter body (and ordinary matter brain). Alternatively this could be a case of excellent clairvoyance on the part of the medium. We saw in chapter 4, in connection with numerous SYNEXs that the range of pure clairvoyance could be considerable, and that people could, by means of good clairvoyance, have access to the inside of books (and their writing or print) and other documents or the written or typed etc., contents of sealed letters. Nevertheless, clairvoyance depends on sampling the correct data in the correct places, whereas a surviving SMM(brain) incorporates these data in a single system, and does not have to search for them in the environment. It is for this reason that I consider this case and similar ones as pointing to genuine survival. The intentionality of the ostensible communicator also points to a surviving entity (see also Case 65 p. 155).

Case 72

The following case, mentioned by Murphy (1945a) was reported by Myers (1903, p. 471), who refers to it as the *Péréliguine Case*. It concerns a seance held on November 18th 1887 in the house of M. Nartzeff, at Tambof Russia. Myers wrote:

The sitting began at 10pm at a table placed in the middle of the room, by the light of a night-light placed at the mantelpiece. All doors closed. The left hand of each was placed on the right hand of his neighbour, and each foot touched the neighbour's foot, so that during the whole of the sitting all hands and feet were under control; [although many sceptics would argue that this type of 'control,' if you wish to call it this, is highly suspect]. Sharp raps were heard in the floor, and afterwards in the wall and ceiling, after which the blows sounded immediately in the middle of the table, as if some one had struck it from above with a fist; and with such violence, and so often that the table trembled the whole time.

M. Nartzeff asked 'Can you answer rationally, giving three raps for yes, one for no?' 'Yes.' 'Do you wish to answer by using the alphabet?' 'Yes.' 'Spell your name.' The alphabet was repeated and the letters indicated by three raps - 'Anastasie Péréliguine.' 'I beg you to say now why you have come and what you desire.' 'I am a wretched woman. Pray for me. Yesterday, during the day, I died at the hospital. The day before yesterday I poisoned myself with matches.' 'Give us some details about yourself. How old are you? Give a rap for each year.' Seventeen raps. 'Who were you?' 'I was a house-maid. I poisoned myself with matches.' Why did you poison yourself?' 'I will not say. I will say nothing more.'

Despite the melodramatic trappings of this case and the hocus pocus that makes it suspect, let us continue. On p. 473 of Myers (1903) is the written statement of Dr

Th. Sundblatt, who wrote:

> On the 16th of this month I was on duty at the local hospital and on that day two patients were admitted to the hospital, who had poisoned themselves with phosphorus... The second of these was a servant Anastasie Péréliguine, aged seventeen, was taken in at 10pm. This second patient had swallowed besides an infusion of boxes of matches, a glass of kerosene, and at the time of her admission was already very ill. She died at 1pm on the 17th and the post-mortem examination has been made today [19th November 1887]... Péréliguine did not state the reason for poisoning herself.

These seance revelations, particularly by the absurdly poorly controlled methods used, and the likely deceptions involved, could be due to somebody at the hospital having told one or more of the seance participants, in advance of the séance and that this person, or persons temporarily escaped from the control, without those 'controlling' him or her daring to state this. Alternatively there could have been telepathy from Dr Sundblatt and/or others who knew of the case history.[7] At any rate if this case were the only one claiming putative survival of a human personality - and even much better cases are, as we have seen, ambiguous (in that they can usually be interpreted in terms of either survival or telepathy or clairvoyance or both combined) it would not be worth considering as crucial or serious evidence.

5.4 The 'Ear of Dionysius' Case

This interesting case, which I shall discuss at some length, claims that two, conceivably 'surviving,' communicators, namely Dr A. W. Verall and Professor Henry Butcher, communicated via a sensitive, an automatic writer, capable of trance. The sensitive, known by the pseudonym Mrs Willett, whose real name was Mrs Combe-Tenant, was married to a brother-in-law of the distinguished parapsychologist and scholar F. W. H. Myers, one of the cofounders of the Society for Psychical Research in London, The case was reported by the Right Honourable Gerald W. Balfour (1917).

On 26 August 1910, Mrs Verall (herself a classics scholar and widow of Dr A. W. Verall) had a session with Mrs Willett, who communicated the phrase *'Dionysius' Ear the lobe,'* among other material, the name Dionysius being pronounced as in Italian. This reference came out of context, unexpectedly, and was meaningless to Mrs Verall at that time. Balfour explains that 'the Ear of Dionysius is a kind of grotto hewn in the solid rock of Syracuse and opening on one of the stone quarries which served as a place of captivity for the Athenian prisoners of war who fell into the hands of the victorious Syracusans after the failure of the famous siege so graphically described by Thucydides. A few years later the quarries were again used as prisons by the elder Dionysius Tyrant of Syracuse. The grotto... has the peculiar acoustic properties of a whispering gallery, and is traditionally believed to have been constructed or utilized by the Tyrant in order to overhear, himself unseen, the conversations of his prisoners. Partly for this reason, and partly from a fancied resemblance to the interior of a donkey's ear, it came to be called *L'Orecchio di Dionisio* or the Ear of Dionysius; but the name only dates from the sixteenth century... The Ear of Dionysius was not referred to again in other Willett scripts until the 10th of January 1914, when the following statements were made by Mrs Willett, who referred to

'a place where slaves were kept - and audition belongs also to acoustics.'

'Think of the Whispering Gallery'

'To toil, a slave, the Tyrant - and it was called Orecchio - that's near.'

'One Ear. . . a one eared place - You did not know (or remember) about it when it came up in conversation, and I said Well what is the use of a classical education.'

'Where were the fields of Enna'

'an ear-ly pipe could be heard'

'To sail for Syracuse'

'Who beat the loud-sounding wave, who smote the moving furros'

'The heel of the boot'

'Dy Dy and then you think of Diana Dimorphism'

'To fly to find Euripides'

'not the Pauline Philemon'

'This sort of thing is more difficult to do than it looked.'

Although this set of statements was directly addressed to Mrs Verall, she was not present at that session, and Balfour suggests that 'the communication must be taken as purporting to come from Dr A. W. Verall.' After Mrs Verall saw the preceding statements she noted on 19th January 1914:

'My typed note on the Willett Script of 16 August 1910 is as follows: "Dionysius' Ear lobe" is unintelligible to me. A. N. Verall says it is the name of a place at Syracuse where D could overhear conversations. This makes clear what was instantly recalled to me on hearing the Willett Script on January 10, that I did not know, or had forgotten what the Ear of Dionysius was, and that I asked AWV to explain it. I cannot say whether on that occasion he asked 'what is the use of a classical education?' but he expressed considerable surprise at my ignorance, and the phrase of the script recalls. . . similar remarks of his on like occasions.'

'The incident to me is very striking. I am quite sure that Mrs Willett was not present when I asked A. W. Verall about the Ear of Dionysius; no one was present except ANV and myself. . . She therefore had no reason to suppose that on this particular subject, of the Ear of Dionysius, my information had been obtained from AWV. . . '

At this stage Balfour comments that he recalls that he and Mrs Willett had previously had a conversation on the Ear of Dionysius, although Mrs Willett could not recall it. Hence, it follows that the 'message' could have derived from Mrs Willett's SMM(brain) without aid from a 'surviving' A. W. Verall, although perhaps not all of it could come from her. 'Dy Dy' may be an attempt to pronounce the name Dionysius. Balfour states that 'the meadows of Enna, a town in Sicily were famous in antiquity as the scene of the Rape of Proserpine.' In all probability Mrs Willett did not know about Enna. Balfour also remarks that, apart from the association of Enna with Sicily there occurs in the script another association 'and a strange far-fetched one. . . in the next line:

'An ear-ly pipe could be heard.'

'The allusion here is apparently to the lines of Tennyson's well-known poem
Tears Idle Tears
The earliest pipe of half-awaken's birds
To dying ears.'
Here the symbol 'ears' seems to be related to the 'ear' in Ear of Dionysius.
Balfour notes that the next reference in the script is almost certainly to the ill-fated Athenian expedition against Syracuse. The words 'who beat the loud-sounding wave, who smote the moving furros' are probably reminiscent of Tennyson's *Ulysses:*
'*Sitting well in order smite*
The sounding furrows,'
Although Balfour thinks that the allusion was not intended. 'The heel of the boot' in the Willett script 'may be taken to indicate the route followed by the Athenian fleet, which passed from Corcyra to Tarentum in the heel of Italy, thence coasted along the toe, and so reached Sicily.'

The preceding comments on the Willett scripts illustrate the subtlety and scholarship behind these communications, a subtlety almost certainly beyond the capacity of Mrs Willett, but well within reach of Dr Verall. Nevertheless, it cannot be excluded that Mrs Willett could have derived her information and structuring telepathically from the (assumed) SMM(brain) of Mrs Verall or one or more other living scholars. I shall not give further extracts from the very long additional case history of the scripts as presented by Balfour. Most of their contents are unlikely to have originated from Mrs Willett herself and, at the very least would require telepathic information from one or more living SMM(brain)s together with highly original construction in association with these SMM(brain)s. Alternatively, these intricate constructions could have originated in the surviving SMM(brain) of the late Dr Verall and Professor Butcher. Balfour states (work cited p. 205) that all the allusions, cited above, might have been within Mrs Willett's knowledge, although I think, that the originality of construction makes this an open question. In any case, Balfour does not make similar claims about the subsequent Willett scripts, that refer to the Ear of Dionysius, in the remainder of his paper, which I shall not discuss here in detail. On p. 209 of Balfour's (1917) paper there occurs in one of the scripts a reference suggesting that two people who were very close friends had thought out the communications, and the literary allusions to them which are very detailed, suggesting that they were the late Dr A. W. Verall of Cambridge University and the other, the late Professor Butcher, 'was himself a highly distinguished Cambridge man, and in later life represented his University in the House of Commons; but he was also for many years Professor of Greek at Edinburgh.'

It might be asked why, if such relatively intricate communications can be synthesized and transmitted to a trance medium (possibly by telepathy from ostensibly surviving SMM(brain)s to the medium's SMM(brain)) are other communications from ostensibly surviving SMM(brain)s to possible SMM(brain)s of trance mediums often of such poor quality? Such communications often do not resemble in style, quality of contents etc., the personality of the supposed communicator, but resemble more the capacities of the medium. This, however, is not difficult to explain. The trance medium's (assumed) SMM(brain) could lack the appropriate endowment of memory traces in order to match the more sophisticated 'messages' from either a living SMM(brain) or a surviving SMM(brain). Likewise, the medium's SMM(brain) may be devoid of sufficiently complex cognitive machinery in order to deal with intricate intellectual input communications. After all, this situation could be analogous to that existing in any university classrooms, where not so well endowed students simply cannot follow the dialogue of a professor or lecturer.

5.5 A materialist view versus the mentalist view of survival

According to the author's point of view it is possible that people's SMM(body)s (including their SMM(brain)s survive death of their ordinary matter bodies; and with the SMM(brain)s there could survive their memories, their capacity to form new memories, their capacity to think creatively and their intentionality. (Intentionality is widely regarded by many philosophers of mind as the hallmark of mentality.) How long such SMM(brain)s could survive stably in the world of Shadow Matter - possibly indefinitely - remains an open question until theoretical Physics tells us a good deal more about Shadow Matter than at present[9] (possibly as deductions from a general Theory of Everything, see p. 23). The present view that the conceivable survival of personality and mentality is vested in the survival of physical systems, namely SMM(body)s and/or their SMM(brain)s, could help to explain the apparent coherence of personal identity which manifests itself in alleged communications from surviving personalities. By contrast, mentalists, like the late Oxford Professor of Philosophy H. H. Price (1953), have, and had, different ideas. Price suggested that people might survive as non-physical disembodied systems of 'mental images,' without any material basis, it would seem. Yet, what keeps this system of mental images together, i.e. associated, in a coherent manner remains unexplained by Price and others who held similar views (e.g. Ducasse, 1951), nor is it clear how 'mental images' could interact with each other. Nobody denies that there exist 'mental images,' but, according to the present theory, they occur as epiphenomenal byproducts of physical activities of the SMM(brain).

Price believed (at least in 1953) that the surviving mental images generate a dream-like world whose dynamics is not that of the physical world, but obeys the 'laws' of Freudian psychodynamics (which I do not subscribe to). With due respect for Price, the findings of Moody (1976) and Sabom (1982) and others, cited earlier, which related to 'out of the body experiences' (OBEs) in near death states, indicate that OBEs are not in the least dream-like, but very realistic, and, in fact, those who had OBEs emphasize the unforgettable sense of reality during OBEs. Thus, if the mental experiences of surviving SMM(brain)s should be similar to OBEs, then there is nothing dream-like about these experiences. In fact, if the hypothesis that OBEs result from largely, or completely out of the body SMM(body)s (and their SMM(brain)s), then one could expect that if SMM(brain)s survive the death of ordinary matter bodies, then surviving SMM(brain)s could have experiences similar to SMM(brain)s during OBEs. If so, then, by hypothesis, surviving SMM(brain)s, like SMM(brain)s in OBEs, could continue to perceive objects and situations in the ordinary world in a manner that resembles ordinary visual perception, although auditory perception seems to be of a telepathic kind in OBEs. Tactile experiences seem absent, although we cannot rule out that there could be tactile experiences resulting from two different SMM(body)s touching each other in the Shadow Matter world. Yet, the important thing is that during OBEs people continue to think normally and to judge situations, recognize other people, objects, etc., in their surroundings, remember things, form new memory traces, and do not live in a dream-like world. Also in an OBE people's 'quasi-material body' (see p. 29) seems to be able to walk along ordinary matter floors over ordinary matter steps, along ordinary matter streets (see Case 5, p. 45). This suggests that the same could apply to a surviving SMM(body).

Whereas a living personality and its personal identity is apparently bound to a particular ordinary matter body, it could, in fact, be bound to, or be a product of, a SMM(brain) (and SMM(body)) and via that SMM(body) and its SMM(brain) could be bound to the ordinary matter body. Accordingly, each person can be associated with a particular ordinary matter body while alive. Likewise, a deceased personality could be associated with just one particular (surviving) SMM(body) (and its SMM(brain)), which survives in the world of Shadow Matter. This surviving SMM(body), apart from interacting with the world of ordinary matter, could also interact with material systems in the world of Shadow Matter. If there should be survival, then I envisage that mutual recognition of two (or more) deceased people could occur partly via telepathy between SMM(brain)s. Also, just as ordinary people can recognize other people from photographs which do not resemble structurally their ordinary matter bodies, but which model people's appearance (in two dimensions) in such a way that their presumed SMM(brain) could (via photons) recognize the people whom the photographs represent, so something similar could happen. A SMM(body) could model in terms of Shadow Matter the ordinary matter body of its owner in such a way that another SMM(body) of another person could, via sphotons and Shadow Matter eyes and the SMM(brain) of that person, recognize the first person (and vice versa).

If there should exist a hereafter, then the objects and people, etc., in it could be constituted by the world of Shadow Matter. Although such a world of Shadow Matter would be consistent with the world of ordinary matter superstrings (see Kolb *et al.* (1985) and section 1.1), there remain, as we have seen, several possibilities for the manner in which SMM(body)s could communicate with each other (e.g. via sphotons that act on Shadow Matter eyes, or communication via SMM(complem)s).

I conclude that my Shadow Matter theory of psi-phenomena would be consistent with much of the evidence cited in this chapter. The theory could be used, at least jointly with other evidence cited earlier, to make out a formidable case for the existence and feasibility of telepathy and clairvoyance and for their possible mechanistic materialistic interpretation. Alternatively some of the preceding case histories of this chapter, when taken jointly with my mechanistic materialistic interpretations of 'out of the body experiences' (OBEs), given in earlier chapters, are consistent with complete SMM(body) (including SMM(brain)) detachment from the dead ordinary matter body of people and are consistent with the survival of the SMM(body) (and its SMM(brain)). Survival of the SMM(brain) is consistent with survival of the human personality after death of a person's ordinary matter body and destruction of that body either by decay or cremation or otherwise.

5.6 Liquidation of alleged difficulties confronting the survival hypothesis

Murphy (1945b) listed various supposed difficulties which some people presumed invalidated the survival hypothesis. First consider the question 'what survives?' The present theory answers this question by hypothesizing that people's SMM(body)s (including their SMM(brain)s) could survive the death of their ordinary matter body. With the surviving SMM(brain) there could survive the human personality, since, according to the theory the SMM(brain) is the 'seat' and 'carrier' of the human personality and of all its psychological functions. I have suggested also already how people could recognize each other's surviving SMM(body)s after death (see p. 167).

It has been asked 'whether the personalities that appear in the forms of phantasms or communicators (in mediumistic trance) actually think and behave in harmony with known traits of the personalities they purport to represent (Murphy, 1945b p. 87).' In the case of a trance communication from an allegedly surviving personality it would be wrong to assume that the communication must necessarily resemble in style that of the familiar style of the deceased. First, it is possible that the communicator cannot reveal himself (or herself), fully, telepathically, via the SMM(brain) of the trance medium, but has to rely on the style of communication which can be produced (according to my theory) by the combined ordinary matter brain and SMM(brain) machinery of the medium. Secondly, it is possible that when SMM(brain)s become severed from the ordinary matter body to which they are attached that they acquire cognitive facilities and modes of expression which they did not possess while bound to the ordinary matter body. In fact, one could expect that surviving SMM(brain)s like body-bound SMM(brain)s can develop further, and, for instance, form new memories and new thoughts. That this is possibly the case is suggested (according to my theory) by OBEs of people who survive near-death experiences and who remember precisely what happened to them while out of the body and what they thought then, suggesting that new memory traces and thoughts were formed by their SMM(brain) while 'out of the body.' The occurrence of OBEs strongly suggests the survival of SMM(body)s (including SMM(brain)s) and the survival of the human personality after death.

One undoubted complication which can always cast doubt on the survivalist interpretation of certain ostensive survival material (but not on OBEs) is the possibility that the communications etc., concerned are based on material picked up telepathically from the living or clairvoyantly picked up from written or printed material contained in letters, manuscripts, books etc., or from objects in the world. If, as my theory assumes, people's SMM(brain)s are constantly bombarded by myriads of SMM(complem)s which are either directly emitted from SMM(object)s or SMM(brain)s or are replicas of such SMM(complem)s, then it remains only for the SMM(brain)s to select from the glut of SMM(complem)s received and to construct out of this appropriate Shadow Matter representations of cognitive constructs with, or without, the help of ordinary matter brains. More problematic still, if survival should genuinely exist, in addition to telepathy and clairvoyance, is that many of the assumed SMM(complem)s that could bombard the SMM(brain)s of the living might not only come from SMM(object)s of inanimate objects and from SMM(brain)s of living people, but they might also derive from surviving SMM(brain)s. This could lead to confusions, say on the parts of mediums, between telepathically picked up material derived from living people, and telepathically picked up material derived from surviving SMM(brain)s. That such confusions might occur is possibly suggested by examples (see Murphy, 1945, pp. 75-6), where the medium seems to think that the communicator is dead, when in reality he or she is living or even in cases where the communicator is fictitiously invented by the sitter.

Another familiar difficulty, raised by many scientists against the survival hypothesis, is the argument that every form of mentality manifested by 'normal' behaviour of people depends on intact ordinary matter brains and nervous systems. It is therefore assumed by many people that in the absence of a normally functioning nervous system, as happens after death, no more manifestations of mentality are possible. This, however, depends on the presumption that we consist exclusively of ordinary matter. This assumption, however, is repudiated by the present theory, which asserts that we are composed of ordinary matter and Shadow Matter and that the SMM(brain) is the producer and seat of mentality. I argued in section 1.1 that contemporary theoretical physics, notably superstring theory, makes the assumption of Shadow Matter plausible. In addition I assumed that ordinary matter and Shadow Matter can form mutually bound states but can also become detached. Those who repudiate the possibility of survival on the grounds that organisms can only consist of ordinary matter, belong to a long line of people with preconceived ideas, who, at one time, included many people who believed that hypnotic states are impossible (see section 1.11). Some who believe on religious grounds in an eternal life of an *immaterial* soul would deny, on a priori premises, that such survival could have a physical basis in Shadow Matter.

It is, I believe, not out of place, in this context, to cite the words of the late Professor Theodore Flournoy, previously quoted by McDougall (1944, p. 507) in a similar context:

It goes without saying that, in order to occupy oneself with the supranormal, it is necessary to admit theoretically the possibility thereof, or what comes to the same thing, to be sceptical of the infallibility and the perfection of science as it now exists. If I consider it *a priori* absolutely impossible that an individual should know, long before the arrival of any telegram, of the accident which has just killed his brother on the other side of the world, or that another person can voluntarily move an object at a distance without the use of a thread in a manner inexplicable by the known laws of mechanics and physiology, then it is clear that I shall raise my shoulders at every recital of telepathy, and shall not stir a step to take part in a séance with the most celebrated of mediums. Excellent means these for enlarging one's horizon and discovering novelties, to recline upon a completed science and a forgone conclusion, entirely convinced that the universe comes to an end at the opposite wall, and that nothing can exist or occur outside that system of daily routine which we have become accustomed to regard as marking the limits of the Real! That philosophy of the ostrich - illustrated formerly by the grotesque pedants at whom Galileo knew not whether to laugh or weep, who refused to put an eye to his telescope for fear of seeing things that had no official right to exist. . . that philosophy is still entertained by many brains petrified by intemperate reading of works of popularized science and by unintelligent attendance at university lectures, those two great intellectual dangers of our time.

Although this passage was written many years ago, it is as valid now as it was then.

5.7 The fallacies of Professor Dodds

I believe that all of Dodds's (1934) often cited objections against the possibility of human survival after death fall flat against the present mechanistic materialistic theory, which, in principle, allows the possibility of the survival of a SMM(body) and its SMM(brain). Dodds (1934 p. 153) wrote:

Anyone who has lived much in the society of the aged and has observed them closely, will I think agree, whatever interpretation he may put upon the facts, that not only the human organism but the human mind, or that portion of it which expresses itself in thought, feeling and action, does appear to grow old. I am thinking here not merely of the gross effects of time, those which a physician would classify as symptoms of senile decay; but also of the subtler psychological changes which come with advancing years, the gradually increasing imperviousness to new ideas, the gradually diminishing response to emotional stimuli, above all, the growing sense of finality, of fulfillment of a destiny accomplished and accepted - in a word, the progressive encroachment upon the will to live of a new will to cease from living. Analogous changes appear, so far as we can judge, to attend the onset of old age in the higher animals, notably the dog and the horse.

That aging affects the ordinary matter body and its ordinary matter brain cannot be disputed, and this is what Dodds was concerned with. On the other hand it remains possible that the SMM(brain) does not age after development. The fact that there may be an estimated daily death of some 50,000 nerve cells of the adult ordinary matter human brain, does not imply any corresponding progressive deficits of the assumed SMM(brain), although increasing impairment of the ordinary matter brain or lesions of that brain could progressively, or suddenly, impair communications (via gravitons) between the ordinary matter brain and SMM(brain). Moreover, if, as I have assumed, in OBEs there occurs a separation of the ordinary matter brain from the SMM(brain), then the reports that in OBEs people. . . began to think more lucidly and rapidly than in physical existence (see p. 43), suggests that the SMM(brain) may be, at all times of normal life, impeded by the ordinary matter brain, and more so in old age, but that this state of affairs does not necessarily exclude the possibility of survival of the SMM(brain). If the SMM(brain) becomes at death liberated from the ordinary matter brain, then the SMM(brain) activity could revert to maximum capacity.

Dodds (1934) also argued that:

Mr Saltmarsh in his valuable analysis of 142 sittings with Mrs Warren Elliott (*Proceedings of the Society for Psychical Research,* vol. 39, pp. 91 ff) found that in this series the total number of veridical statements about events subsequent to the death of the supposed communicator was actually larger than the total number of such statements about events prior to his death; and that the percentage of veridicality was also higher in the statements about events subsequent to death.

Dodds believes that this speaks against the possibility of survival, but it does nothing of the sort. For instance, in my kind of theory, if the SMM(brain) of a deceased person were to survive death of the ordinary matter body, then the SMM(brain) could, just as it can, by hypothesis, in an out of the body experience, form new memories and continue to think, etc. It could form new memories, partly with the help of telepathy and clairvoyance partly with the help of the Shadow Matter eyes, etc., of the possibly surviving SMM(body) to which the surviving SMM(brain) could belong, and it could form new memories of new thoughts. Thus, events occurring after death of the ordinary matter body of a person could be recorded by the surviving SMM(brain) of that person.

There are several further questions sometimes raised by skeptics (e.g. Dodds, 1934) concerning the survival hypothesis. Where precisely do surviving entities, if there are any, reside, and how do they spend their time? Why, if those, who were brilliant intellects on Earth, do they not, if surviving as such, communicate any profound statements (but see section 5.4 for a possible exception)?

In a different context the originators of contemporary Shadow Matter theory, Kolb *et al.* (1985, p. 419) concluded in the journal *Nature* (London) that

The effect of Shadow Matter is hard to detect in everyday life; the reader could be living in the middle of a shadow mountain or at the bottom of a shadow ocean. But it would have many effects in the early and contemporary Universe...

The questions, raised above, could be answered as follows. If SMM(body)s and their SMM(brain)s should, by any chance, survive, then they could possibly reside among us, although we cannot see them any more than we can see the possible 'out of the body' SMM(body)s of people who undergo 'out of the body experiences(OBEs),' since we cannot literally see the Shadow Matter world and its objects. Moreover, the lack of profound 'communications' via trance mediums from deceased outstanding intellects does not necessarily speak against the possibility of SMM(brain) survival. It could mean simply that the means of communication (e.g. the medium's existing memory traces), which would have to be used to express these ideas, are inadequate (and that, say, sufficiently appreciative intellects may, as a rule, not be mediumistically gifted). A deceased high calibre mathematician, like John von Neumann or an earlier one, like David Hilbert, could hardly communicate any subtle new mathematical theorems which they had discovered with their SMM(brain), since the death of their ordinary matter body, via a sensitive medium who lacks the most rudimentary memory traces with which to symbolize the new theorems. In fact, we cannot exclude the possibility that some sudden inspirations of men of genius could be due to unknown telepathy between them and the SMM(brain)s of deceased people of genius.

It must be stressed, once more, that if there should occur communication between surviving SMM(brain)s of people whose ordinary matter body has died and SMM(brain)s of living people, then such communication is here assumed to be based on the mechanism of telepathy suggested earlier (see p. 85). If this hypothesis of telepathic communication from surviving agencies (i.e. SMM(brain)s) is valid, then it would be very difficult to argue that the telepathic hypothesis, whereby messages coming ostensibly from the dead come by telepathy from the living (SMM(brain)s) is to be preferred to the view that these messages come telepathically from surviving SMM(brain)s. (I do not think that Dodds (1934) realized this possibility, since he did not know of Shadow Matter theory and the Shadow Matter mechanisms proposed in this book.) In fact, it has been shown that some mediumistic messages, ostensibly coming from the dead seem to have come by telepathy from living people (Dodds, 1934). Yet, this does not rule out that we receive also, via mediums, or directly (by telepathy), information from surviving SMM(brain)s.

Significantly, Dodds (1934 p. 162) wrote:

. . . Dr Osty tells us that one of his sensitives, Mlle de Berly, is capable of 'reproducing approximately the timbre and rhythm' of voices of persons living or dead whom she has never heard speak, of 'saying what they might say or might have said' and of 'exhibiting their customary attitudes' (Osty, *La Connaisance Supranormale*).

To accomplish such a feat by telepathy from either living or surviving SMM(brain) would be consistent with my theoretical model which assumes that SMM(brain)s of living or surviving people could represent not only those parts of ordinary matter brains that deal with particular memories but represent the whole of the speech region and other motor parts of ordinary matter brains in terms of Shadow Matter. Thus, by telepathy the speech-representing part of a living, or, alternatively, of a surviving SMM(brain) could communicate messages from its engrams (i.e. memory traces) to the speech representing part of the SMM(brain) which is bonded, by hypothesis, to the ordinary matter brain of the medium (or appropriate other sensitive person).

As regards the way in which surviving SMM(brain)s could spend their time, the answer could be that, like living SMM(brain)s, they meditate a lot, ranging from daydreaming to serious thought, and they could, via the Shadow Matter eyes of the possibly surviving SMM(body) also recognize the ordinary world and its objects (i.e. the Shadow Matter representation of that world and its objects), although, except by telepathy, they can no longer interfere with that world and its objects (including people). They can read but not write.

When it comes to object reading (see sections 2.5), Dodds (1934, p. 164) referred to this as 'puzzling occurrences on any hypothesis,' a view that I did not share in section 2.5, where I proposed mechanisms for explaining object reading (popularly known as psychometry) in mechanistic terms within my framework of Shadow Matter theory. By contrast, Dodds (1934, p. 164) shared some of the views of the great embryologist Hans Driesch (who was an unrepentant vitalist) and stated that 'I can only echo Driesch's remark that object-reading is "the most mysterious and inexplicable of all supernormal phenomena,"' an opinion here repudiated.

Notes to Chapter 5

1. 1977, e.g. p. 5 and p. 9

2. This sense of 'clarity' was noted in numerous OBEs as a class characteristic by Moody (1976) (See the 'striking sense of reality' section 1.6 p. 37

3. Also, since M'Connell was involved in the accident at about the time of his

'appearance' to Larkin, he could not possibly have been near Larkin, so that we are dealing with a genuine hallucination.

4. This intentionality of an ostensibly surviving personality seems typical of several possible survival cases in this book e.g. **66**, **69-71**.

5. There is also CSICOP, mentioned earlier, an organized group of people engaged on a crusade to destroy the credibility of parapsychology. Such people often think that by suppressing the truth about psi-phenomena they make the world more rational, that they create a saner world by calling black white and white black, that they are admirable people by being standard bearers of 'the truth,' the 'truth' being what they believe in and not necessarily what is true.

6. The 'bright light' phenomenon is also typical of some OBEs (see Moody, 1976)

7. Murphy (1945a, p. 15) states in a footnote that 'for other cases in which information is given, ostensibly by a dead person, concerning facts known to the deceased but unknown to any living person, see *Proceedings of the Society for Psychical Research* (London) vol.17, pp. 181-2, vol.35, pp. 511 ff, and vol.36, pp, 303-5.'

8. Murphy (1945a, p. 17) stressed that in this and other similar cases (reported in the *Journal of the Society for Psychical Research* (January-February 1940, pp. 142-52) it is always conceivable that somebody gleaned information normally or paranormally from newspapers or other sources.

9. For indefinite survival the second law of thermodynamics may not have to apply to Shadow Matter.

6
Precognition and Shadow Matter and the Possible Self-Assembly of Superintelligences

6.1 A Dark Story in the History of Parapsychology

When it comes to precognition, we must remember that there was a time from, say, the mid 1940s to, perhaps, the mid 1970s, when many parapsychologists hailed the statistically unprecedented highly significant series of card guessing experiments conducted by Dr Samuel Soal. These experiments allegedly demonstrated, as was widely believed, the occurrence of precognitive telepathy. On the strength of these ingeniously designed experiments Soal received a Doctorate of Science (DSc) from the University of London, and was made a Senior Lecturer in Mathematics at Queen Mary College, University of London, and Soal was also made president of the Society for Psychical Research (London). Some sceptics, however, were more cautious. They argued that nobody, apart from Soal, had ever obtained results of remotely comparable statistical significance (in the technical sense) in the whole of parapsychology, and so Soal's work became suspect among certain parapsychologists. Finally, it became completely discredited when Betty Markwick (1978), with the aid of fast working computers (which Soal could not have anticipated in 1942), discovered serious data manipulation in Soal's work.

It could be argued that Soal stood to gain a DSc and the financial rewards of a Senior Lectureship, as a result of his manipulations. Yet this fraud-implying interpretation of Soal's work is not the only possibility. There exists an important alternative. Soal had been my teacher at Queen Mary College, and he, in fact, had drawn my attention to parapsychology. Later I corresponded a good deal with Soal. He believed obsessively that by all means fair or foul he must outmatch the American parapsychologist Professor J. B. Rhine and put British experimental parapsychology firmly on the map. So it could be that Soal's fraud was not done for personal gain, but (as he may have thought!) for the sake of enhancing British standing in the subject. If so, it was not obvious to Soal at that time that the opposite could occur when high speed computers would come into existence. As a scientist Soal did not show that great scientific intuition, which marked out Professor Herbert Fröhlich, FRS, whose Research Assistant I was for a time at the H. H. Wills Laboratory at Bristol University, where I worked in theoretical physics. Fröhlich, with whom I was pretty friendly in those days, died recently (see his obituary in *The Times* (London) 30th January, 1991, p. 16). He was not only a man of great intuition in theoretical physics, and highly original and of international status, but was, at least in recent years, of the opinion that parapsychology is an important subject. There will always be people who, in my opinion, are less far sighted than the late Herbert Fröhlich, and who, like Professor Max Hammerton (1983), formerly Professor of Psychology in the University of Newcastle upon Tyne, will spare no effort to denigrate parapsychology. Hammerton, and other hypersceptics certainly know some of the facts of parapsychology, but they believe, dogmatically that these facts can invariably be explained in terms of fraud, trickery or gullibility and so forth. Undoubtedly there have been some fraudulent mediums, and there was Dr Soal. Yet, the possibility that there could be new physical insights that could explain psychic phenomena has simply not dawned on individuals like Hammerton, who, together with some of less academic standing (e.g. Randi, 1991) make a feast out of debunking parapsychology, without succeeding.

6.2 Some Case Histories of Spontaneous Precognition

Compared to the number of good case histories of apparitions (see Tyrrell, 1953) and putative evidence relating to survival (which may alternatively be interpreted as evidence for telepathy and/or clairvoyance, as appropriate), the number of good, and well corroborated, case histories of spontaneous precognitions is small. The books by Tyrrell (1948), Celia Green (1976) and various other writers contain a few, perhaps impressive, case histories. There exist also at least two, perhaps more, collections of case histories of precognitions (Sidgwick, 1888-9; Saltmarsh, 1934). At least some of these cases have been well corroborated, and the credibility of others seems plausible.

The following case is taken from Sidgwick's (1888-9) paper (p. 292).

Case 73

The case is from Mrs Alger, who at the time she wrote in January 1883, lived at Hedsor Lodge, Belmont, Twickenham, SW. She writes:

'Some years ago in March my husband, who is an army tutor, asked me to call at the Civil Service offices for some papers. I had come from Victoria Station, walking towards the Abbey, when just before crossing over to Canon Row, I felt someone touch me on the shoulder. I turned round and saw my husband's mother, looking very death-like. I said, "Oh mother, what a start you gave me!" but she had gone. A feeling of great depression came over me, and I was quite unable to go on my husband's errand, but went home. All the way home I thought of what had happened, and as I got indoors I made up my mind to tell my husband and then at once go to Brixton, where his mother lived. However, I fainted before I saw Mr Alger, and after recovering, I felt unwell, so that I had to go to bed. After thinking the matter over I said nothing of what I had seen, but early in the evening when my husband came into my room, I asked him to go and see his mother. We were talking it over as to whether it would be right to leave the boys by themselves, when I heard a voice say, "Come both of you on the 22nd" (the 22nd of March is my birthday). I at once told my husband my day's experience and added, "My birthday will be your mother's death-day." Mr Alger went at once to Brixton, and on his return told me his mother had a cold, but was, on the whole, as well as ever; but on the 22nd of March, that is four days after, we stood at her death bed.'

Alger confirmed most of this account, although he did not state that the day foretold by the apparition was the 22nd of March. So all that can be claimed is that, at least in a general way, and probably in a very specific way, Mrs Alger had foreknowledge of the death of her mother-in-law in the near future, and at a time when there were no good reasons for thinking that she would die so soon.

Case 74

The following case is also taken from Mrs Sidgwick's (1888-9) paper (p. 305). It is from Mrs Morrison of 131 Cornwall Road, Westbourne Park, London. The incident occurred in Province Wellesley, Straits Settlements in the East Indies in May 1878 and an account of it was sent to the Society for Psychical Research (London) in 1882 Mrs Morrison writes:

And last of all a sweet little girl, the pride of its parents' heart was taken. Some days prior to the child's illness, I was lying awake one morning when I distinctly heard a voice say, 'If there is darkness at the 11th hour there will be death.' In alarm I started up in bed and the same words were slowly and deliberately repeated.

Naturally enough, when about a week after, the child was taken seriously ill, I watched with perturbed feelings and grave anxiety the aspect of the sky day and night, the moon being at full just then. Two or three days passed; the little one hovered between life and death; above the sun blazed with unmitigated fervour, relentless heat, no sign of cloud or disturbance of the atmosphere in any way. Twice in the course of every 24 hours was 11 o'clock looked for with trembling apprehension. At last after more than a week of this cloudless weather and a few minutes before 11 in the morning a squall arose with extraordinary suddenness, servants flew to close the Venetian shutters, making the inside of the house extremely dark. The sky became black with clouds, and my heart sank. That day soon after one o'clock the child's spirit quitted its little mortal frame. . . I cannot be mistaken as to the time when the darkness came, as I had to consult my watch a little time before in order to give the child medicine.

Mrs Morrison's reference to the child's spirit, is, of course, not relevant to the mechanistic type of theory developed in this book, but I had to report it as stated. If foreknowledge is indeed involved in this case, then this suggests that the physical state of the atmosphere in any locality on Earth, together with the preconditions that can change that state, must be fore-knowable to within minutes or seconds, and similar remarks would have to apply to the physical states of countless objects, animate and inanimate in the universe. A possible theoretical interpretation of how this could come about within the present theory will be discussed later.

The following case comes also from Mrs Sidgwick's (1888-9) collection (p. 311):

Case 75

The lady who communicated the case (in 1882) to the Society for Psychical Research (London) wished to remain anonymous. She was taken to a trance medium, and writes, amongst other things:

> Though I had only arrived in Boston the day before, her guides instantly recognized that I came over the water, and opened up, not only my past life, but a great deal of the future. They said I had a picture of my family with me, and on producing it, the medium told me (in trance) that two of my children were in the spirit world, and pointing to one son in the group, she said 'You will soon have this one there; he will die suddenly, but you must not weep for him. . .'
>
> I had not been home many weeks, before my son, a brave boy of 17, was killed at a game of football.

Again, I disregard the spiritualistic innuéndos by the medium. Also, if this was a case of foreknowledge, then, once more, myriads of physical states of myriads of ordinary material systems must be foreknowable in principle.

Next, I shall present a case cited by Mrs Sidgwick (1888-9, pp. 321 ff). She begins 'for clearness' with the account of the death involved taken from the *York Herald*.

Case 76

York Herald, Friday, July 28, 1882:

'SCARBRO. SAD DEATH OF A GENTLEMAN VISITOR - An accident of a melancholy character, and which unfortunately has been attended with fatal results, occurred on Wednesday evening to a London gentleman named Frederick Schweizer, who for the past few days has been staying at the Grand Hotel. It appears that on the afternoon of that day the deceased, along with a casual acquaintance named Deverell, who stayed at the Castle Hotel, went for a ride on horseback along the beautiful Forge Valley rides. When near Ayton the deceased was somewhat in advance of his companion, and it is surmised that his steed shied at the white gate; anyhow he was thrown on to the road, and the horse galloped away. His friend on getting up to him dismounted, and a passing carriage was utilized to convey him to his hotel. This was at six o'clock, and three hours subsequently the deceased expired, it is supposed from concussion of the brain.' The accident occurred on July 26th.

Mr Schweizer's mother, Mrs Schweizer, now of 6, Addison Road North, wrote on October 28th 1882: (presumably to the Society for Psychical Research, London):

'I send you the particulars of the dream I had just eight days before it was

177

realized, though why I could not be told of the unfortunate accident as it occurred I can't understand, nor why Henry Irving's name should be mentioned. - J. Schweizer.

On the 18th of July I had the following dream or vision (I can't say which) - I was walking on the edge of a high cliff, the open sea in front, dear Fred and a stranger a little in advance, when Fred slipped suddenly down the side of the cliff, and in doing so gazed with the most intense anguish into my very soul. I shall never forget that look. I turned to the stranger and said, "May I ask who you are and what is your name?" He replied, "My name is Henry Irving." I said, "Do you mean Irving the actor?" He said, "No, not exactly, but something after that style." I said in my reply, "Now that I look at you, you have the same agonized expression in your face that I have so often noticed in Irving's photographs in the shop windows." So I awoke in a miserable state of mind. It was between 5am and 6am. The servants came down soon after. The dream seemed to haunt me; I could think of nothing else. When I met my eldest son John, at breakfast, I asked at once, where was Fred? (I must state here that Fred was a travelling partner of three brothers, and then in the north of England on a journey.) His brother, after hearing the dream, said, 'Oh, Fred is alright; he is in Manchester. . . " '

A few days later John said to Mrs S 'Fred says he is going to take a week's holiday at Scarborough,' and Mrs S objected to this and wanted Fred to come home and was going to write to him. She continues 'On this day (when she had the argument with John about Fred's return) before I got out of bed at the same hour, between 5 and 6am, a person seemed to pass the side of my bed, and said into my ear in an audible voice, "You are not done with trouble yet." ' Mrs S 'started up and awoke and related the matter at breakfast while talking of Fred to John, and said, "I think it was your father." He said, "Oh nonsense that is like the dream you bothered me about a few days ago." This was on the 23rd. On the 24th John mentioned, that he had had a telegram from Fred to send on £10, that he was enjoying himself immensely, that the weather was glorious. . . In the morning of the 26th I went to the letter-box and found a telegram for John, which announced an accident to Fred. John, however, did not like to tell me, and hurried off to the office. I asked John the nature of the telegram, but he said 'business'. On arriving at his office, there was a telegram of a similar kind from the hotel proprietor at Scarborough. Poor Fred was dead at the time, as he only survived the accident three hours. John and I set off at once and found all over, and next day it was proposed that we should visit the fatal spot. His companion in the unfortunate excursion accompanied us. He sat opposite to me in the carriage, and when I looked at him I remembered the dream of the 18th, and recognized the stranger who had the agonized expression and asked him at once if his name were 'Henry'. He said "yes my name is Henry," when I told the dream. He then said, "The most extraordinary part is, I am connected with the Volunteers, and we have private theatricals, and I recite, and am always on those occasions introduced as Henry Irving, jun."'

(Mrs Schweizer says that an account written by her and substantially the same as the above, was signed by her son and by Mr Deverell.) Mr Deverell was not acquainted with Frederick Schweizer and was only introduced to him on the afternoon of the accident, and Mr Deverell was not known directly or indirectly until then to any member of Mrs S's family. Significantly 'Mrs Schweizer did not see horses in her vision' (see Sidgwick, 1888-9, p. 323). Hence, a precognition need not exactly resemble the things and the situations that are being precognized.

The next case from Mrs Sidgwick's (1888-9) collection comes from Mr James Cox, Admiralty House, Queenstown, Ireland (Secretary ret. to the Admiral commanding in Ireland). It is dated December 18th 1883.

Case 77

On Sunday 11th September 1881, while proceeding in *H. M. S. Phoenix*, from Newfoundland to Halifax, Nova Scotia, I dreamt that one of my brother officers was lying dead in a house in Portsmouth. The dream was so vivid that it quite disturbed my mind the following morning, and it was with difficulty that I could shake off the uncomfortable feeling. At breakfast I sat opposite the officer, and looking round the table, I remarked: 'I dreamt last night that I saw one of you fellows lying dead, but I won't say which, as I don't want to spoil your appetite.' In the course of the afternoon as we were steaming into Halifax harbour, the officer was sitting at the stove in the wardroom joining in an animated conversation about the speed of the ship, etc. A few minutes after we anchored, I went on shore, and returned again on board at 10pm and as I was about to go below to my cabin, the officer of the watch motioned me to be silent, and approaching me, said 'Poor S, is dead, he just died suddenly'; and as I passed across the mess-room I beheld the officer of my dream lying dead in his cabin.'

The subject of the dream was Mr Sharp, chief engineer of the *Phoenix*, who died suddenly of heart disease, and the account of Mr Cox was confirmed by a brother officer, whose statement is printed in Mrs Sidgwick's paper (p. 331).

The following case was contributed by the secretary of the Munich *Psychologische Gesellschaft* to *Sphinx* for March 1887 (Sidgwick 1888-9, pp. 335 ff). It states amongst other things:

Case 78

In a night early in August 1886, I was witness, in a dream, to the outbreak of a rapidly spreading conflagration, which through its terrifying grandeur had a paralyzing effect on me. When I woke I remained so under the influence of what I had dreamed, that the reality of such misfortune could not have distressed me more. Strange to say, soon after waking the thought pressed upon me that our securities, which the brewery-proprietor B. kept in a fire-proof safe, were in danger. Although I cannot remember having dreamed of any danger to the bonds, and though there was no external reason for connecting the paper with the fire, to my astonishment despite all the reasons with which I endeavoured to talk myself out of this apparently motiveless feeling, the idea increased to such a point, that I at once told those around me about my dream. As though my misgiving was to be confirmed as correct, three days later I had exactly the same dream, only with still greater distinctness. The unaccountable uneasiness increased still more, and I had the sensation as though an internal voice called on me, to put the bonds to safety. As the loss of them would have meant a great misfortune for us, I tried (following the warning) to induce my husband to put the papers in some other place.

As the majority of persons in his place would probably have done, he looked upon my fears as groundless and could not attribute any importance to a dream. At first he flatly refused to grant my request. But in the meanwhile the inexplicable feeling of anxiety so thoroughly took possession of me that I made him continually more urgent representations. At last, after about ten days, he gave in, less on account of the dream than for the sake of my comfort. From the moment that I knew that the bonds had been placed in security, in the Munich Mortgage and Exchange Bank, my equanimity was restored. Soon afterwards I went into the country, to the Tyrol, and should hardly have thought more of this occurrence, had I not suddenly during the night of the 14th-15th of September, again been the dreaming witness of a tremendous fire. But instead of, as before, being frightened by the exciting scene, there came over me a feeling of relief as of being saved from great calamity, by the timely saving of the papers. On the morning of the 15th I made known my dream experience to those around me. Sadly enough the warning was fulfilled, for already, the following day, I received written information that the brewery in which was the above mentioned safe, had been reduced to ashes by a destructive fire, which broke out on the 14th of September. As I afterwards heard, the building was burnt to the ground; the fire-proof safe was exposed to flames and heat for 36 hours, so that the proprietor's papers which were preserved in it were completely charred. . .

The correctness and precision of this communication is confirmed by the signature of five witnesses. The husband of the lady testifies in his protocol that he really had been led to the removal of the papers by her request, as above described; he was also a witness of the breaking out of the fire on the 14th September. Three friends of the family, Frau von O., Herr von M., and Baron von E., state in their evidence that the above dream had been fully related to them during the first days of August, and that they themselves had taken part in the discussion respecting the danger of the bonds. Further Herr von M. states that before the arrival of the letter on the morning of the 15th of September, in the Tyrol, the dream of the previous night (also given above) had been related to him. According to the statement of the thus heavily visited proprietor of the brewery, the fire broke out on the 14th of September, thus some four weeks after delivering-up the bonds, and raged three days; the 36 hours of heat which the safe had sustained had destroyed all papers that were in it. Moreover several newspapers lie before me which give an account of this great conflagration.

Although the first and second dream were, presumably, precognitive, the third dream could be due to clairvoyance or telepathy.

Many other interesting cases of apparent foreknowledge are given in comparable detail by the Sidgwick report, and will not be cited here. Instead I shall give some cases cited by Tyrrell (1948). The first of these cases is taken from the *Journal of the Society* *for Psychical Research,*vol. 1, p. 283.

Case 79

Tyrrell (1948, pp. 74) summarizes the case history thus:

The narrator of the. . . case was personally interviewed by Sir William F. Barrett FRS. In January 1887, Captain A. B. Macgowan, an officer in the American army, was on leave at Brooklyn with his two boys, then on vacation from school. 'I promised the boys,' he said, 'that I would take them to the theatre that night and I engaged seats for us three. At the same time I had the opportunity to examine the interior of the theatre. I went over it carefully, stage and all. Those seats were engaged the previous day, but on the day of the proposed visit it seemed as if a voice within me was constantly saying "Do not go to the theatre; take the boys back to school." I could not keep these words out of my mind; they grew stronger and stronger and at noon I told my friends and the boys that we would not go to the theatre. My friends remonstrated with me, and said I was cruel to deprive the boys of a promised and unfamiliar pleasure to which they had looked forward, and I partly relented. But all afternoon the words kept repeating themselves and impressing themselves upon me. That evening, less than an hour before the doors opened, I insisted on the boys going to New York with me and spending the night at a hotel convenient to the railroad by which we could start in the early morning. I felt ashamed of the feeling that impelled me to act thus, but there seemed no escape from it, That night the theatre was destroyed by fire with a loss of some three hundred lives.'

In conversation with Sir William Barrett, Captain MacGowan said that the voice was perfectly clear, 'like someone talking inside me, it kept saying: "Take the boys home, take the boys home".' And this from breakfast time till he took the boys away, shortly before the theatre opened. He had never experienced anything like it before or since; never had any other hallucination. . . Three hundred and five people were burnt to death that night.

Tyrrell (1948) cited another case which also seems to involve a hallucinatory warning by an inner voice, as in **Case 79**:

Case 80

This case comes from the *Journal of the Society for Psychical Research* (London) vol 8, p. 45

The account first-hand by the principal witness, was published eighteen years after the event. It is corroborated by the landlady, by the witness's husband, and by the Miss W. whose life was saved. Both the last two asserted that their memory of the event was perfectly clear.

In July 1860 the wife of Rev Dr W. with her little daughter A (names known) and a servant went to stay at Trinity, near Edinburgh. One Sunday afternoon the child went alone to play on a strip of ground between the railway and the sea wall, which was enclosed by a gate at each end. On this fine summer's day nothing apparently could have been safer. Soon after the child had gone, Mrs W reports that she distinctly heard a voice, as it were, within me, say, 'Send for her back or something dreadful will happen to her' but reason rebelled and she refused to do so. The same thing recurred, the words being repeated with greater emphasis. Mrs W could only think that the child might possibly meet a mad dog; but she tried to throw off the feeling. Then the words were repeated for the third time, and a feeling of terror seized her. She rang the bell for the maid and sent her out to fetch the child in. Later in the afternoon, an engine ran off the line and crashed through the sea wall on to the rocks where the child admitted that she had been intending to sit. No sooner had the child been withdrawn from the dangerous spot than all feeling of anxiety passed away.

The following case from the *Proceedings of the Society for Psychical Research*, vol 11, pp. 577 ff, is taken from Tyrrell's summary. It is a 'dream-case,' but the dream is recurrent.

Case 81

The husband and step-father of the narrator, Lady Q. (name known), corroborated the case. The main points were noted down soon after the event. As a child, Lady Q, having lost her father, and her mother having married again, went to live with her uncle, to whom she became very attached. She says: 'In the spring of 1882 I dreamed that my sister and I were sitting in my uncle's drawing-room. In my dream it was a brilliant spring day, and from the window we saw quantities of flowers in the garden, many more than were in fact to be seen from that window. But over the garden there lay a thin covering of snow. I knew in my dream that my uncle had been found dead by the side of a certain bridle-path about three miles from the house - a field road where I had often ridden with him, and along which he often rode when going to fish in a neighbouring lake. I knew that his horse was standing by him, and that he was wearing a dark homespun suit of cloth made from the wool of a herd of black sheep which he kept. I knew that his body was being brought home in a wagon with two horses, with hay in the bottom and that we were waiting for his body to arrive. Then in my dream the wagon came to the door; and two men well-known to me - one a gardener, the other a kennel huntsman, helped to carry the body upstairs, which were rather narrow. My uncle was a very tall and heavy man, and in my dream I saw the men carrying him with difficulty, and his left hand hanging down and striking against the banisters as the men mounted the stairs. This detail gave me in my dream an unreasonable horror.'

Lady Q slept no more that night and in the morning looked so changed and ill that her uncle asked what was the matter. She told him and begged him never to go alone by that particular road. Two years after the dream recurred 'with all its details the same as before,' and she told her uncle again. In May 1888, she was in London, now married and expecting a baby, when the dream recurred for the third time. This time it was followed almost at once by her uncle's death. From her step-father and the old nurse she discovered that the details of the death were the same as those of the dream, even to the bruising of the hand as the body was being carried upstairs. She stated that she had only two other impressive dreams in her life. Flowers and snow were for certain reasons symbolical in her family of death; and these seem to have been introduced into the dream to reinforce the death-prediction. . .

I believe that the recurrence of this dream, and other pre-cognitive dreams in other cases, could simply mean that the first occurrence of the dream built up a memory trace (engram) sequence, which subsequently led to recall of the memorized dream.

6.3 A Materialistic Physicalistic Theory of Precognition

In a recent paper (Wassermann, 1988) I wrote:

Some people believe that precognition rules out any conceivable physical theory of psi-phenomena. They ask, 'how can an event which, because it has not yet happened, does not yet exist, be the cause of our knowing about it? And what becomes of human freedom if we can see in advance the action of a person who has not yet decided what he is going to do?' (Tyrrell, 1948, pp. 73-4). I think that 'human freedom' may be an illusion, but a useful one (Wassermann, 1988a).

In order to explain precognition in terms of Shadow Matter theory, I shall argue as follows. According to my theory every 'object' of ordinary matter has an SMM(object) bound to it, and the latter, in turn, can give rise to SMM(complem)s. An 'object' in this context could be an electron, or a quark, say, or it could be an atom of a particular kind, or it could be a specific type of molecule, or it could be a chair, say, or a table, or a house, or the 'object' could be the whole of planet 'Earth' including its atmosphere, and all the objects associated with the planet (including all living creatures), or the 'object' could be the sun, or it could be the whole solar system, and so forth. In other words, there could be object specific SMM(object)s of vastly different sizes and complexities and each SMM(object), and in turn, could generate SMM(complem)s specific for that SMM(object), and these SMM(complem)s could separate from their parent SMM(object) template and could elastically contract or expand, as the case may be.

I now assume that SMM(complem, Earth), i. e. any SMM which is complementary to the SMM which is bound to the Earth (and all its animate and inanimate matter) can contract so as to become so small that it can interact with the SMM(brain) of people. I also assume that SMM(complem, Earth) can provide by means of its Shadow Matter an advanced representation (i.e. an advanced modelling) of states of the ordinary matter world of the Earth before these ordinary matter states occur in the ordinary matter world. (The mechanism envisaged is, thus, in its major respects, similar to the mechanisms already postulated for telepathy and clairvoyance.) By means of its advanced modelling a contracting and miniaturized SMM(complem, Earth) could then, by interacting with SMM(brain) generate foreknowledge. This model assumes, of course, that the modelling by SMM(complem Earth) of states in the ordinary matter world may take place at a much faster rate, or tempo, than the rate of occurrence of the states in the ordinary matter world. Thus, what precedes causes of ordinary matter events on Earth are not ordinary matter events that are caused by these 'causes,' but 'causes' in the Shadow Matter world, namely Shadow Matter causes which lead to the advanced modelling (or advanced representation) of ordinary matter events on Earth by SMM(complem, Earth). Thus, physics, and with it materialism (i. e. materialism that includes Shadow Matter) could be consistent with the occurrence of foreknowledge.

That in foreknowledge people become aware of a representation of the future has also been recognized before by other writers (e. g. Tyrrell, 1948, p. 91), although these writers did not suggest a suitable mechanism and machinery whereby such representations could be established, nor the type of material systems (i. e. Shadow Matter) which could enact such representations and establish them. In my type of theory the 'creator' of an advanced representation is SMM(complem, Earth), and by acting on a SMM(brain) of some person it could produce the foreknowledge.

Tyrrell (1948, p. 91) who did not operate in terms of such a mechanistic model, noted, nevertheless, that:

These *ad hoc* creations do not slavishly copy the events they stand for; they are often inaccurate, apparently for two reasons. (1) Because the representative creation is dramatic in character, and the creator uses artistic licence or frank symbolism. (2) Because the representative creation is often multiply caused, the factor relating to the future is adjusted so as to fit with the rest of the theme...

I believe that Tyrrell's way of putting things is somewhat obscure, simply because he seemed to lack a mechanistic model, and, hence, fell back on mentalism, a kind of philosophy that is not to my liking.

6.4 Liquidation of an Elementary Fallacy

One critic who saw an earlier version of the present theory argued that if SMM(brain)s were to exist on the Earth surface, then they would drop straight to the centre of the Earth under the Earth's gravitational pull. This argument, however, disregards the possibility envisaged here that to all ordinary matter objects in the universe there are bound corresponding SMM(object)s. In particular I postulated above that every ordinary material constituent of the Earth is bound to a corresponding Shadow Matter constituent, so that the whole Earth is bound to a Shadow Matter SMM(Earth). Just as the ordinary matter on the Earth and in the interior of the Earth prevents people (and their brains) who populate the Earth surface from falling towards the centre of the Earth, despite gravitational attraction, so the surface structures and 'interior' of the Shadow Matter SMM(Earth) could prevent the SMM(body)s (and their SMM(brain)s) from dropping to the centre of SMM(Earth).

6.5 The Possibility of the Self-Assembly of Super-intelligences

It would be arrogant to assume that we must necessarily be the highest intelligences in the universe. In fact, those who believe in deities uncompromisingly reject this assumption. Yet, one can also question it if one is a materialist, say a mechanistic materialist like the present author. It could be assumed, for instance, that SMM(brain)s of different people could become assembled as super-SMM(brain)s. This, while it cannot be ruled out in advance, is however, questionable, since each SMM(brain) is assumed to exist only (like any SMM(object)) in a single copy within the universe, so that an encounter of suitable SMM(brain)s of different people, after the death of these

people, is likely to be a very rare event. By contrast the assumption that SMM(complem, brain)s of different SMM(brain)s (whether of living or of dead people) could combine and form, by self-assembly, complexes of SMM(complem, brain)s seems more promising. SMM(complem, brain)s exist, by assumptions made earlier in this book, in many copies, so that at least one copy of a particular SMM(complem, brain) could encounter at least another copy of a different SMM(complem, brain) and combine with it. In a similar way larger, multi-SMM(complem, brain) complexes, consisting of many SMM(complem, brain)s could be formed, and, by replication, give rise to many copies of such complexes.

Such multi-SMM-complexes could be capable of more intricate intellectual achievements than individual SMM(complem, brain)s, because they could share the different memory traces, and other existing cognitive structures, of a possibly great variety of different SMM(complem, brain)s. This would give us, perhaps, something more explicit, in mechanistic materialistic terms, than C. G. Jung's 'collective unconscious,' but would, possibly, not be too far removed from Jung's notion, although the latter may strike mechanistic materialists as a little nebulous. Jung often saw something important (e.g. the synchronicity experiences discussed earlier) but then lapsed into mysticism in order to explain it (e.g. the 'acausal connecting principle' suggested as an 'explanation' of synchronicity experiences and psi-phenomena). Of course, many people believe that it is irrational to accept the genuineness of many alleged kinds of paranormal phenomena. But what should be questioned is not the rationality of beliefs in psi-phenomena but the rationality of the belief that psi-phenomena are irrational. With some luck the theory presented in this book may help to convince those who are not hopelessly prejudiced that psi-phenomena are capable of rational explanations within mechanistic materialism.[1]

Notes to Chapter 6

1. I have intentionally abstained from discussing so-called physical psi-phenomena, such as spoon beading, psychokinesis, poltergeists and other claims to move matter in the world of ordinary matter by paranormal means. Some of these claims, notably by stage entertainers, and others, that spoon bending is due to paranormal means, have been vigorously contested, and the evidence for poltergeists, although not insubstantial, is very restricted. The claims of certain stage entertainers to produce spoon bending at will suggests that they are not producing a typical psi-phenomenon, because of the spontaneity of typical psi-phenomena. In fact the evidence for genuine, spontaneous 'physical' psi-phenomena with repeatable class character-istics seems very slight, with the possible exception of poltergeists (see Wolman, 1977a pp. 382 ff). But see the Epilogue.

Epilogue:
Recent
Psychokinesis
Experiments
at
Cambridge University
(England)

There have been many experiments to demonstrate psychokinesis (PK), a few of them on an impressive scale. PK is the alleged faculty whereby human beings (and possibly animals as well) can produce 'psychological sets' (i.e. attitudes) which result in particular influences on material systems outside them. Since, according to my theory such 'psychological sets' for PK are generated by the SMM(brain), it may be assumed that SMM(complem)s emitted by the SMM(brain) and replica of these SMM(complem)s could reach, randomly, targets for which the SMM(complem)s are 'set'. By means of emission of streams of gravitons the SMM(complem)s could then have specific effects on the 'chosen' material targets.

Let me just cite some recent experiments which seem to demonstrate PK and which are summarized lucidly by Neville Hodgkinson and Maurice Chittenden in *The Sunday Times* (London) 5 January, 1992, p. 4. They write, among other things:

Good news for gamblers: the winning streak is all in the mind. Scientists believe psychokinetic powers can influence the spin of the roulette wheel or the fall of dice.

An international conference in Athens yesterday was presented with reports on machines devised to generate and measure random electronic impulses that seem to have been influenced by the willpower of bystanders. When left to themselves, the machines flash a row of 31 lights running either to the left or right, in a random way.

However, the scientists have found that some people without any contact with the machine can 'will' the lights to flash significantly more in one direction than the other. Nobody succeeds all the time, but experiments at Cambridge

University's Cavendish laboratory have shown the phenomenon occurring again and again.

'It does seem the mind can influence random events without any connection to the physical world,' said Dr Fotini Viras, the physicist in charge of the tests. She believes the effect can be seen in the winning streak, the feeling that occasionally comes to gamblers that fate is making everything go their way. Dr Helmut Schmidt, a senior research associate at the Mind Science Foundation in San Antonio, Texas, who devised the test machine, said: 'if you really want it to work, it does.'

There exist also reports of the occurrence of large scale physical effects possibly related to the PK mechanisms. These phenomena known as 'Poltergeists' are rare and spontaneous and seem to happen in the presence of a particular person who seems to be causatively involved without any physical intervention by that person via that person's normal neuro-muscular system. For surveys of earlier work on experimental PK the reader should consult Stanford (1977) and for Poltergeists see the survey by Roll (1977), although I am apt to be on the side of the hypersceptics when it comes to evaluations of some of the Poltergeist phenomena.

We are now approaching the end of this novel and far-reaching theory of psi-phenomena. These are likely to have existed since the beginning of human evolution and conceivably, since the beginning of evolution of life on Earth, and psychic phenomena might occur in all animal species. At least this would be compatible with the assumed all-pervading presence of Shadow Matter. I have shown how telepathy, clairvoyance and 'out-of-the-body-experiences' as well as apparitions and precognition could all be explained extensively in terms of the interaction of Shadow Matter and ordinary matter. Synchronicity experience (SYNEXs), in turn, could be explained in terms of psychic phenomena, notably clairvoyance. Many of my explanations could be extended also to other species. We have also seen that survival of the Shadow Matter body (and its Shadow Matter brain) after death of the ordinary matter body could be consistent with known facts. So we have arrived at the end of a comprehensive theory, apparently the first of its kind, that enables us to explain psychic phenomena by linking them with theories of mainstream science, notably super-string theory and Shadow Matter theory. Some hypersceptics might argue that Shadow Matter is, as yet, an entirely theoretical construct and has not been observed. So how can psychic phenomena be explained in terms of something hitherto not directly observed? There exists, however, an apparent parallel (and possibly several more) in the history of science.

Perhaps the best known is the postulate by Planck in 1900 of light quanta, which were then as unobserved as Shadow matter is nowadays. By means of light quanta Planck explained, and derived ingeniously, a new formula for black body radiation which agreed excellently with observations and replaced the previous, inadequate formula, propounded by other physicists and derived from classical statistical thermodynamics. For this work Planck obtained later a Nobel Prize. However, as Messiah (p. 10)[1] put it:

Upon its publication, Planck's hypothesis seemed unacceptable; physicists almost unanimously refused to see therein more than a lucky mathematical artifice which could, some day, be explained within the framework of classical doctrine. The very success of Planck's theory could not be considered as irrefutable proof that the energy exchanges between matter and radiation on the microscopic scale actually take place by quanta... One could likewise doubt the validity of the quantum hypothesis itself, in the same manner as one had doubted for a long time the validity of the atomic hypothesis for lack of being able to verify it directly on the microscopic level...

Just as I have argued that psi-phenomena confirm the hypothesis of Shadow Matter and are, at present, the best indirect evidence for its existence, whereas the properties of Shadow Matter could explain most or all psychic phenomena, so we have a parallel situation when we return to Planck's quantum hypothesis. In 1905, in a celebrated paper, for which he obtained later a Nobel Prize, Einstein explained the photoelectric effect, previously experimentally established, but not explained, by P von Lenard in 1902. Einstein's explanation relied on Planck's quantum hypothesis, and gave strong confirmation to the latter. Thus Planck's hypothesis and Einstein's explanation are closely interlocked. Similarly the Shadow Matter hypothesis and the present explanation of psychic phenomena can be seen to be intimately related. I conclude that material entities such as quanta or Shadow Matter, which either could not or cannot be directly observed, can, nevertheless, form the basis of extensive explanations of a wide range of phenomena. Typically, Shadow Matter can, apparently, link parapsychology with physics and, thereby, with a central branch of science. In this way physics may advance towards its possible ultimate theory, namely a **Theory of Everything.**

Notes to Epilogue

1. Albert Messiah (1965) *Quantum Mechanics,* vol 1, translated from the French by G. M. Temmer, Amsterdam, (North-Holland Publishing Co)

Appendix

Neurological evidence consistent with the hypothesis of a Shadow Matter brain

There are large numbers of reported cases where much of the normally present ordinary matter brain is congenitally absent, although this, apparently does not affect mental functions. If, as I have assumed, mental functions are vested in the Shadow Matter brain, rather than in the ordinary matter brain, then absence of much of the ordinary matter brain need not interfere with the autonomous function of the Shadow Matter brain, which could have developed normally, being guided in its development by the residual parts of the ordinary matter brain.

Consistent with this J. Lorber (1981, 'The disposable Cortex,' *Psychology Today,* vol.15, p. 126) wrote:

I have now studied more than 100 people who are among the most severe hydrocephalic cases, using a CAT scan [a computerized brain scanning technique], formal intelligence tests, and detailed neurological assessments. The majority show normal physical development as well as good intelligence. Among them are several with IQs of 130 or more who are doing splendidly at school. Some of the adults have university degrees, others are members of executive committees, one is a senior nurse, and most are in gainful occupations. They hear and see well. One young man has a first-class honours degree in mathematics at a university, despite the fact that some 95% of the cranial areas that normally accommodate the hemispheres are filled with fluid. Instead of the normal 4.5cm thickness of brain tissue between the ventricles and the surface of the cerebral cortex, he has a layer measuring a millimetre or two.

Yet, it is known that cases of damage to normal brains can result in serious mental defects. In such cases, however, most or all of the Shadow Matter brain could remain firmly linked to the ordinary matter brain, including parts of the ordinary matter brain that are damaged; and do not function properly any more. This residual linkage, combined with partial malfunctioning of the ordinary matter brain, could then prevent the Shadow matter brain from functioning properly.

References

Allport. G. W. (1954) *The Nature of Prejudice*, Reading, Mass., Addison-Wesley.

Assailly. A. (1963) 'Psychophysiological correlates of mediumistic faculties,' *International Journal of Parapsychology, 5*, 357-74.

Backman, A. (1892) 'Experiments in clairvoyance,' *Proceedings of the Society for Psychical Research (London), 7*, 199-220.

Balfour. G. W. (1917) 'The Ear of Dionysius,' *Proceedings of the Society for Psychical Research* (London), *29*, 197-243.

Beloff, J. (1970) 'Parapsychology and its Neighbours,' *Journal of Parapsychology, 34*, 129-42.

Beloff, J. (1977) 'Historical Overview,' in *Handbook of Parapsychology* (B.B. Wolman ed.) pp. 3-24 New York, Van Nostrand Reinhold.

Beloff, J. (1979) 'Could there be a physical explanation for psi?' In *Abstracts of the Third International Conference of the Society for Psychical Research*, Edinburgh, p. 1, Society For Psychical Research, 1 Adam and Eve Mews, London W8 6UG.

Beloff. J. (1987) 'Parapsychology and the Mind-body Problem,' *Inquiry, 30*, 215-25.

Bergson, H. (1911) *Matter and Memory* (translated from the French by N. M. Paul and W. Scott Palmer) London, George Allen and Unwin.

Bishop, P. O. (1970) 'Beginning of form, perception and binocular depth discrimination in cortex,' in *The Neurosciences, Second Study Program* (F. O. Schmitt, editor-in-chief) pp. 471-85, New York, Rockefeller University Press.

Bliss, T. V. P. (1990) 'Memory: Maintenance is presynaptic,' *Nature* (London), *346*, 698-99.

Brain. W. R. (1951) *Mind, Perception and Science*, Oxford, Blackwell.

Clowes, J. S. and Wassermann, G. D. (1984) 'Genetic control theory of Developmental events,' *Bulletin of Mathematical Biology, 46*, 785-825.

Crick, F. C. (1984) 'Neurobiology, memory and molecular turnover,' *Nature* (London), *312*, 101.

Dodds, E. R. (1934) 'Why I do not believe in Survival,' *Proceedings of the Society for Psychical Research*, (London) *42*, 147-72.

Douglas, A. (1976) *Extra-Sensory Powers*, London, Victor Gollancz Ltd.

Ducasse, C. J. (1951) *Nature, Mind and Death*, La Salle, Illinois, Open Court Publishing Co.

Duff, E. G. (1986) *Fundamental Particles* An introduction to Quarks and Leptons, London, Taylor and Francis.

Eccles, J. C. (1986) 'Do mental events cause neural events analogously to the probability-fields of quantum mechanics,' *Proceedings of the Royal Society*, London B *227*, 411-28.

Ellis, J. (1986) 'The superstring theory of everything, or nothing?' *Nature* (London) *323*, 595-98.

Ellis, J. (1987) 'Strings in four dimensions,' *Nature* (London) *329*, 488-89.

Eysenck, H. J. (1953) *Uses and Abuses of Psychology*, Harmondsworth, Penguin.

Flew, A. G. N. (1953) *A New Approach to Psychical Research*, London, Watts and Co.

Gauld, A. (1977) 'Discarnate survival,' In *Handbook of Parapsychology* (B. B. Wolman ed.) pp. 577-630, New York, van Nostrand Reinhold Co.

Gibson, J. J. (1950) *The Perception of the Visual World*, Boston, Mass., Houghton Mifflin.

Gindes, B. C. (1953) *New Concepts of Hypnosis*, London, George Allen and Unwin.

Goddard, G. V. (1986) 'A step nearer a neural substrate,' *Nature* (London) *319*, 721-22.

Goelet, P., Castelluci, V. F., Schachar, S. and Kandel, E. R. (1986) 'The long and the short of long-term memory - a molecular framework,' *Nature* (London) *322*, 419-22.

Green, C. (1968) *Out-of-the-body-experiences*, London, Hamish Hamilton.

Green, C. (1976) *The Decline and Fall of Science*, London, Hamish Hamilton.

Green, C. and McCreery, C. (1975) *Apparitions*, London, Hamish Hamilton.

Green, M. B. (1985) 'Unification of forces and particles in superstring theories,' *Nature* (London) *314*, 409-14.

Green, M. B. and Schwarz, J. H. (1984) 'Anomaly cancellations in supersymmetric D=10 gauge theory and superstring theory,' *Physics Letters, 149* B, 117-22.

Green, M. B., Schwarz, J. H. and Witten (1987) *Superstring Theory,* 2 vols. Cambridge University Press.

Gurney, E., Myers, F. W. H. and Podmore, F. (1886) *Phantasms of the Living*, 2 Vols, London, Society for Psychical Research (abridged edition 1918, London, Kegan Paul.).

Hammerton, M. (1983) Book Review of *Explaining the unexplained: Mysteries of the Paranormal* by H. J. Eysenck and C. Sargent, *Quarterly Journal of Experimental Psychology, 35A*, 550.

Hansel, C. E. M. (1980) *ESP and Parapsychology: A Critical Reevaluation*, Buffalo, New York 14215, Prometheus Books.

Hardy, A., Harvie, R., and Koestler, A. (1973) *The Challenge of Chance* London, Hutchinson.

Hart, H. and Hart, E. B. (1933) 'Visions and apparitions collectively and reciprocally perceived,' *Proceedings of the Society for Psychical Research* (London) *41*, 205-249.

Huxley, J. S. (1942) *Evolution: The Modern Synthesis*, London, Allen and Unwin.

Julesz, B, (1971) *Foundations of Cyclopean Perception*, Chicago University Press.

Julesz, B. (1975) 'Experiments in the visual perception of texture,' *Scientific American, 232* April 34-43.

Jung, C. G. (1955) In *The Interpretation of Nature and Psyche*, by C. G. Jung and W. Pauli, London, Routledge and Kegan Paul.

Kalmus, P. T. (1983) 'High-energy physics W and Z particles from CERN,' *Nature* (London) *304*, 686.

Kandel, E. R. and Schwartz, J. H, (1982) 'Molecular biology of learning: modulation of transmitter release,' *Science, 218*, 433-43.

Kant, I. (1900) *Dreams of a Spirit-seer* translated by Emanuel, F. Foerwitz (edited with an introduction and notes by Frank Swell) London, Swan Sonnenschein & Co.

Kling, J. W. and Riggs, L. A. (eds.) (1971) *Woodworth and Schlosberg's Experimental Psychology*, London, Methuen.

Köhler, W, and Wallach, H. (1944) 'Figural aftereffects and investigation of visual processes,' *Proceedings of the American Philosophical Society, 88*, 269-357.

Kolb, E. W. , Seckel, D. and Turner, M. S. (1985) 'The shadow world of superstring theories,' *Nature* (London) *314*, 415-19.

Lashley, K. S. (1950) 'In search of the engram,' In *Symposia of the Society for Experimental Biology 4*, pp. 454 ff.

Leaf, W. (1889-1890) 'A record of observations of certain phenomena of trance: part 2,' *Proceedings of the Society for Psychical Research* (London) *6*, 558-646.

Leonard, G, O. (1931) *My Life in Two Worlds*, London, Cassell.

Levin, B. (1980) *Taking Sides*, London, Pan Books.

Lodge, O. (1916) *Raymond*, London, Methuen.

Long, A. A. (1974) *Hellenistic Philosophy*, London, Duckworth.

McDougall, W. (1944) *An Outline of Abnormal Psychology* (4th ed.) London, Methuen.

Macneile Dixon, W. (1937) *The Human Situation*.

Marks, D. F. (1986) 'Investigating the paranormal,' *Nature* (London) *320*, 119-23.

Markwick, B. (1978) 'The Soal-Goldney experiments with Basil Shackleton; new evidence of data manipulation,' *Proceedings of the Society for Psychical Research* (London) *56*, part 211 pp. 256-77.

Mason, A. A. (1952) 'A case of congenital ichthyosiform erythrodermia of Brocq treated by hypnosis,' *British Medical Journal* 23rd August, 422-23.

Mead, G. R. S. (1919) *The Doctrine of the Subtle Body in Western Tradition*, London, Watkins.

Moody, R. A. Jr. (1976) *Life after Life*, Harrisburg, PA., Stackpole Books.

Morris, R. G. M., Anderson, E., Lynch, G. S. and Baudry, M. (1986) 'Selective impairment of learning and blockade of long-term potentiation by an N-methyl-D-aspartate receptor antagonist AP5,' *Nature* (London) *319*, 774-76.

Morton, R. C. (1892) 'Record of a haunted house,' *Proceedings of the Society for Psychical Research* (London) *8*, 311-32.

Moss, T. (1976) *The Probability of the Impossible*, London, Routledge and Kegan Paul.

Muldoon, S. J. and Carrington, H. (1969) *The Projection of the Astral Body*, New York, Weiser.

Murphy, G. (1945a) 'An outline of survival evidence,' *Journal of the American Society for Psychical Research, 39*, 2-34.

Murphy, G. (1945b) 'Difficulties confronting the survival hypothesis,' *Journal of the American Society for Psychical Research, 39*, 67-94.

Murphy, G. (1945c) 'Field theory and survival,' *Journal of the American Society for Psychical Research 39*, 181-209.

Myers, F. W. H. (1892) 'On indications of continued terenne knowledge on the part of phantasms of the dead,' *Proceedings of the Society for Psychical Research* (London) *8*, 170-252.

Myers, F. W. H. (1903) *Human Personality and its Survival of Bodily Death*, 2 vols. London, Longmans, Green and Co.

Osis, K. and Haraldsson, E. (1977) *At the Hour of Death*, New York, Avon books.

Osty, E. (1923) *Supernormal Faculties of Man*, London, Methuen.

Parker A. (1975) *States of Mind: ESP And Altered States of Consciousness*, London, Malaby Press.

Penfield, W. and Rasmussen, T, (1950) *The Cerebral Cortex of Man*, New York, Macmillan.

Penrose, R. (1976a) 'The Nonlinear Graviton,' *General Relativity and Gravitation, 7*, 171-76.

Penrose, R. (1976b) 'Nonlinear Gravitons and Curved Twistor Theory,' *General Relativity and Gravitation 7*, 31-52.

Piper, A. L. (1929) *The Life and Work of Mrs. Piper*, London, Kegan Paul, Trench, Trubner and Co.

Polani, M. (1951) *The Logic of Liberty*; Reflections and Rejoinders, London, Routledge and Kegan Paul.

Price, H. H. (1953) 'Survival and the idea of "another world",' *Proceedings of the Society for Psychical Research*, (London) 50, 1-25.

Price, G. R. (1955) 'Science and the supernatural,' *Science, 122*, 359-67.

Price, H. (1931) 'The R101 disaster (case record): mediumship of Mrs. Garrett,' *Journal of the American Society for Psychical Research, 25*, 268-79.

Prince, Morton (1906) *The Dissociation of a Personality*, New York, Longmans and Co.

Puthoff, E. H. and Targ, R. (1981) 'Rebuttal of criticisms of remote viewing experiments,' *Nature* (London) *292*, 388.

Radclyffe-Hall, M. and Troubridge, U. (1918-19) 'On a series of sittings with Mrs. Osborne Leonard,' *Proceedings of the Society for Psychical Research* (London) *30*, 339-554.

Randi, J. (1982) *Flim-Flam, Psychics, ESP, Unicorns and other Delusions*, Buffalo, N. Y. Prometheus Books.

Randi, J. (1991) *Psychic Investigator*, London, Boxtree.

Rhine, J. B. (1977) 'Extrasensory perception,' In *Handbook of Parapsychology*, (B. B. Wolman ed.) pp. 163-74, New York, Van Nostrand Reinhold.

Rhine, J. B. and Pratt, J. G. (1957) *Parapsychology: Frontier Science of the Mind*, Springfield Ill., Charles C. Thomas.

Rist, J. H. (1972) *Epicurus*, Cambridge University Press.

Robinson, J. O. (1972) *The Psychology of Visual Illusion*, London, Hutchinson University Library.

Rock, I. (1970) 'Perception from the standpoint of psychology,' *Perception and its Disorders, Research Publications Association Res. Nerv. Ment. Disorders, 48*, 1-11.

Roll, W. G. (1977) 'Poltergeists,' In *Handbook of Parapsychology* (B. B. Wolman ed.) pp. 382-413, New York, Van Nostrand Reinhold Co.

Sabom, M. B. (1982) *Reflections of Death: a Medical Investigation*, London, Harper and Rowe.

Saltmarsh, H. F. (1934) 'Report on Cases of Apparent Precognition,' *Proceedings of the Society for Psychical Research* (London) *42*, 49-103.

Sidgwick, E. M. (Mrs. H) (1888-89) 'On the evidence for premonitions,' *Proceedings of the Society for Psychical Research* (London) *5*, 288-354.

Sidgwick, E. M. (Mrs. H.) (1891)'On the evidence for clairvoyance,' *Proceedings of the Society for Psychical Research* (London) *7*, 30-99.

Sidgwick, E. M. (1920-1921) 'An examination of book tests obtained with Mrs. Leonard,' *Proceedings of the Society for Psychical Research* (London) *31*, 253-60.

Sidgwick, H. (1894) 'Report on the census of hallucinations,' *Proceedings of the Society for Psychical Research*, (London) *10*, 25-422.

Sidis, B. and Goodhart, S. (1905) *Multiple Personality*, An experimental Investigation into the Nature of Human Individuality, New York, D. Appleton and Co.

Spoehr, K. T. and Lehmkuhle, S. W. (1982) *Visual Information Processing*, San Francisco, W. H. Freeman and Co.

Stanford, R. G. (1977) 'Experimental psychokinesis: a review from diverse perspectives,' In

Handbook of Parapsychology, (B. B. Wolman ed.) pp. 324-81, New York, Van Nostrand Reinhold.

Steiner, R. (1969) *Occult Science - An Outline*, London, Rudolf Steiner Press.

Stevens, C. F. (1989) 'Strengthening the synapses,' *Nature* (London) *338*, 460.

Stryer, L. (1988) *Biochemistry* (3rd ed.), San Francisco, W. H. Freeman and Co.

Targ, R. and Puthoff, H. (1974) 'Information transmission under conditions of sensory shielding,' *Nature* (London) *251*, 602-7.

Tart, C. T., Puthoff, H. E. and Targ, R. (1980) 'Information transfer in remote viewing experiments,' *Nature* (London) *284*, 191.

Thouless, R. H. (1984) 'Do we survive bodily death?' *Proceedings of the Society for Psychical Research* (London) *57*, 1-52.

Tyrrell, G. N. M. (1948) *The Personality of Man*, West Drayton Middlesex, Penguin Books.

Tyrrell, G. N. M. (1953) *Apparitions* (The seventh F. W. H. Myers Memorial Lecture of the Society for Psychical Research) (Second revised edition) London, Gerald Duckworth.

Vernon, M. D. (1952) *A Further Study of Visual Perception*, Cambridge University Press.

Vernon, M. D. (1962) *The Psychology of Perception*, Harmondsworth, Penguin.

Vernon, M. D. (1970) *Perception through Experience*, London, Methuen.

Wald, G. (1981) 'The molecular basis of visual excitation,' in *Molecular Processes in Vision* (E. W. Abrahamson and S. E. Ostroy eds.) Stroudsburg, Pennsylvania, Hutchinson Ross Publ. Co.

Wassermann, G. D. (1955) 'Some comments on methods and statements in Parapsychology and other sciences,' *British Journal for the Philosophy of Science*, 6, 122-46.

Wassermann, G. D. (1974) *Brains and Reasoning*, London, Macmillan.

Wassermann, G. D. (1978) *Neurobiological Theory of Psychological Phenomena*, London, Macmillan.

Wassermann, G. D. (1972) *Molecular Control of Cell Differentiation and Morphogenesis*, New York, Marcel Dekker.

Wassermann, G. D. (1973) 'Molecular Genetics and Developmental Biology,' *Narure New Biology 245*, 163-65.

Wassermann, G. D. (1982b) 'TIMA Part 1. TIMA as a paradigm for the evolution of molecular complementarities and macromolecules,' *Journal of Theoretical Biology*, *96*, 77-86.

Wassermann, G. D. (1982c) 'TIMA Part 2. TIMA-based instructive evolution of macromolecules and organs and structures,' *Journal of Theoretical Biology. 99*, 609-28.

Wassermann, G. D. (1983) 'Quantum mechanics and consciousness,' *Nature and System. 5*, 3-16.

Wassermann, G. D. (1986) 'The wiring in of neural nets revisited,' *Bulletin of Mathematical Biology*, *48*, 661-80.

Wassermann, G. D. (1986a) 'Note on the abnormality of the operator for the structural gene of the dopamine D_1 receptor as a possible partial cause of schizophrenia,' *Journal of Theoretical Biology*, *120*, 277-83.

Wassermann, G. D. (1988) 'On a physical (materialistic) theory of psi-phenomena based on Shadow Matter,' *Inquiry 31*, 217-22.

Wassermann, G. D. (1989) 'Theories, systemic models (SYMOs), Laws and Facts in the Sciences,' *Synthese, 79* 489-514.

Wentzel, G. (1949) *Quantum Theory of Fields*, New York, Interscience.

West, D. J, (1954) *Psychical Research Today*, London, Gerald Duckworth.

Whiteman, J. H. M. (1977) 'Parapsychology and physics,' In *Handbook of Parapsychology*, (B. B. Wolman ed.) pp. 730-56, New York, Van Nostrand Reinhold and Co.

Wolman, B. B. (1977a) (editor) *Handbook of Parapsychology*, New York, Van Nostrand Reinhold and Co.

Wolman, B. B. (1977b) 'Mind and body: a contribution to a theory of parapsychological phenomena,' in *Handbook of Parapsychology*, (B. B. Wolman ed.) pp. 861-79, New York, Van Nostrand Reinhold and Co.

Index